WELDING FOR VEHICLE RESTORERS

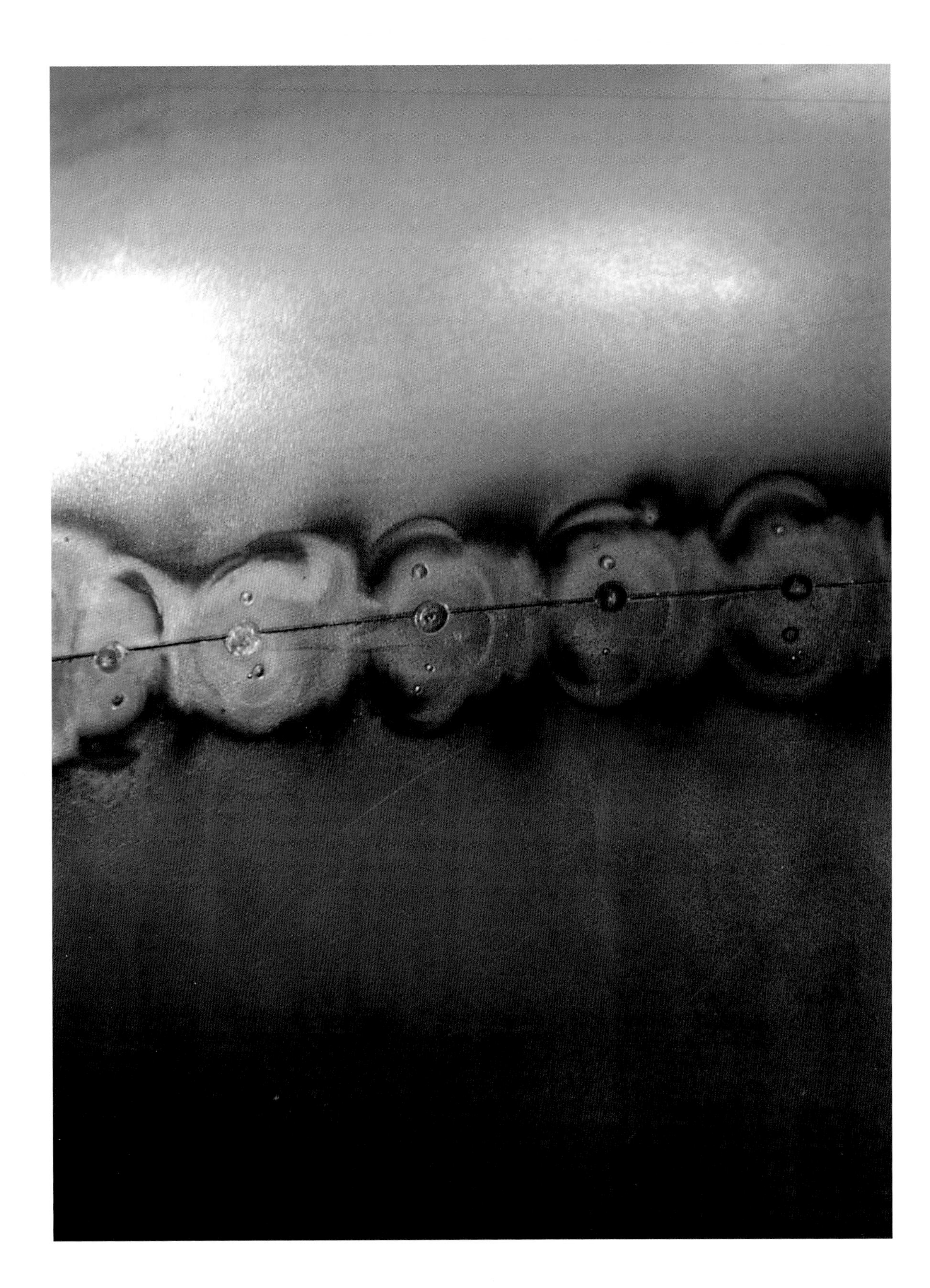

WELDING FOR VEHICLE RESTORERS

Bruce Macleod

THE CROWOOD PRESS

First published in 2020 by
The Crowood Press Ltd
Ramsbury, Marlborough
Wiltshire SN8 2HR

enquiries@crowood.com

www.crowood.com

This impression 2023

British Library Cataloguing-in-Publication Data
A catalogue record for this book is available from the British Library.

ISBN 978 1 78500 681 4

Disclaimer
Safety is of the utmost importance in every aspect of an automotive workshop. The practical procedures and the tools and equipment used in automotive workshops are potentially dangerous. Tools should be used in strict accordance with the manufacturer's recommended procedures and current health and safety regulations. The author and publisher cannot accept responsibility for any accident or injury caused by following the advice given in this book.

Designed and typeset by Guy Croton Publishing Services,
West Malling, Kent

Printed and bound in India by Thomson Press India Ltd.

CONTENTS

DEDICATION

This book is dedicated to the memory of my father, Iain Macleod.

He was one of the last of a generation who trained in the craft of coachbuilding whilst it was a current method of production for bespoke vehicle bodywork in Britain. He was also passionate in passing on these skills to the next generation.

ACKNOWLEDGEMENTS

Thanks to Sarah, my loving wife, for her support and encouragement and her time spent reading through the draft and correcting my poor grammar.

Thank you also to my learned friends Peter Dalrymple and Will Simonson, who took the time to read through the draft and for their suggestions to help it all make sense.

Thanks to Michelle Cooke for her help in correcting errors and taking photographs when needed.

The images used in the book are mostly drawn from my archive as well as various external sources. Special thanks to Steven Booth, who took most of the photographs required to illustrate the techniques. Thanks to R-Tech Welding for supplying the images of some of the equipment shown.

INTRODUCTION

ABOUT THE AUTHOR

I began my career in the late 1970s as an apprentice engineer with a machine tool manufacturer. It was here that I learnt to make things to precise dimensions by hand and using precision machinery.

My main interest was in hand-making things, so following my apprenticeship I went to work for my father, who had trained as a coachbuilder from the age of fifteen. At that time, my father was running his own business restoring and hand-making body panels for the Jaguar XK range of vehicles.

This is where I first learnt to weld on vehicle body panels, restoring original bodyshells as well as making new panels, initially welding mild steel panels before progressing to welding aluminium alloy and stainless-steel panels and more uncommonly used materials such as brass, copper and titanium alloy.

During my career, I restored numerous vehicle bodies, working in a variety of materials and construction methods. I have also built complete new bodies from scratch in mild steel and aluminium alloys, from replica early coach-built alloy-bodied cars to later post-war all-steel bodyshells.

Possessing a comprehensive experience in all aspects of vehicle body construction and repair, as well as qualifications in joinery and tool-making, I became a highly regarded authority in my craft. Through the car body restoration courses that I have been running for over two decades, I have trained countless professionals and amateurs in the making of body panels and the restoration of original vehicle bodies.

The author, Bruce Macleod. STEVEN BOOTH

OUTLINE OF THE BOOK

The text of this book concentrates on the welding of mild steel and aluminium alloy, which are generally used for vehicle body panels, with some reference to the welding of brass and stainless-steel sheet metal used for trim panels on older vehicles.

The result of welding is very different to soldering and brazing. When welding, two pieces of metal are joined by both parts being melted to form one single piece of metal of equal strength. When soldering and brazing, two pieces of metal are joined by the inclu-

Restored 1952 Jaguar XK 120 fixed-head bodyshell.

sion of another metal. This added metal has a lower melting point than the metal being joined together and therefore creates a weaker joint. Welding is therefore the far superior method to use in forming a strong joint between metal parts.

This book covers the various methods of welding used to make and repair sheet-metal body panels. It does not cover the welding of cast materials, as this involves vastly different methods of working and is best done by a specialist, as the knowledge and equipment needed to carry out an effective repair is beyond the means of most vehicle restorers. The welding of higher carbon steels used on more modern vehicle bodies is also not covered. Oxyacetylene welding methods are not included either, as these have been largely superseded by Tungsten Insert Gas (TIG) welding. The welding of thin sheet metal requires different techniques to those used in the welding of thicker metal plate, bar and tube.

I have set out to explain as simply as possible the methods, materials and equipment I have found to be most useful in my career. The process may seem a daunting prospect for the reader, so although this book aims to be a helpful guide to those seeking to weld sheet-metal panels in a professional manner, I cannot overemphasize enough the benefit of attending a hands-on training course to learn these methods. This has proved to be the best way to understand fully the principles and to gain the necessary skills in the shortest time.

Where some forms of welding require minimal skill level, others will require a significant amount of practice to become skilled at carrying out the techniques effectively. When restoring a classic car of some value it is worth putting the time and effort into becoming skilled in the use of all the techniques covered in order to achieve a professional finish to your work. The essential equipment required for this work is minimal, so even the novice welder working from home can achieve a professional standard of finish if using the correct techniques.

There will be a degree of repetition in and among chapters. This is necessary to reinforce certain principles that are critical to the processes.

TYPES OF WELDING USED ON CLASSIC VEHICLES

Early car bodies were of a simple design that had no welded joints. The body panels were either nailed to wooden frames or riveted together to form larger structures.

Oxyacetylene welding and brazing were initially used to join sheet-metal panels. Gas-shielded electric-arc welding and electrical resistance spot welding were developed as more complicated steel and aluminium bodies were produced. The first monocoque, or chassisless, car, which made full use of electrical resistance spot welding in its construction, was developed by Citroën in 1934. Gas-shielded electric-

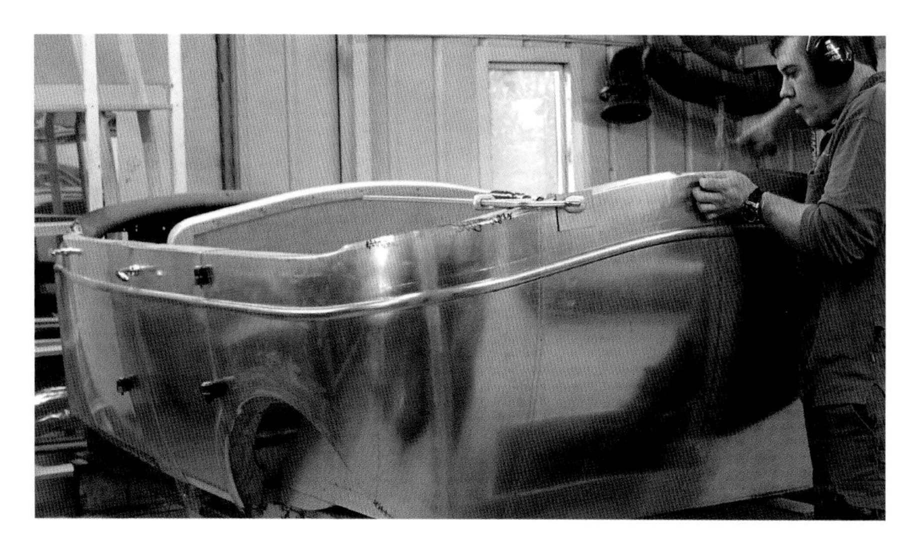

Early coach-built car body consisting of a timber frame, which is panelled in aluminium alloy and sits on a heavy chassis.

Citroën Traction Avant. This was the first monocoque, or unity, body used on a car. The body design made full use of resistance spot welding in its assembly. MARC VORGERS

arc welding in the form of MIG (Metal Inert Gas) and TIG were developed in the 1940s. Spot welding has remained the most prolific means of joining sheet-metal panels due to its simplicity in application and the fact that it requires the least skill in its use.

MIG and TIG welding require a greater level of skill in their operation, which is acquired by practice in the use of the necessary techniques. Welding using these techniques demands patience, good eyesight and quick reactions. Control over the welding torch is critical to the process; stimulants such as caffeine are best avoided as they will impair your ability to control the torch smoothly.

The following chapters cover the tools and techniques required to carry out welding for the restoration of a vehicle body to a professional standard and in the most efficient manner.

CHAPTER ONE

HEALTH AND SAFETY

Health and safety should always be taken seriously, as severe injury and long-term health conditions can result from a lack of awareness and discipline in using protective measures. To reduce the risk of harm to yourself and others, it is always necessary to understand the risks and to use the appropriate protective equipment. It is important to develop safe working habits, as a momentary lapse in concentration when working on vehicles can cause serious injury. Tiredness plays a part in many accidents, so avoid carrying out any operations using power tools or other dangerous equipment when feeling fatigued. Working alone increases the risk of harm, as you are more likely to lift something that is too heavy for one person. You may also be unaware of other incidents in the workshop, such as a fire starting, if you are concentrating on a particular job. The environment you are working in can create risks, as many household garages have poor floor surfaces, leaky roofs and outdated electrics, all of which increase the chances of an accident occurring.

Always perform a quick visual risk assessment before carrying out any work on a vehicle so as to reduce the risk of injury or harm. The appropriate PPE (Personal Protective Equipment) should be used in order to reduce the risk of harm, as there are serious health issues that can arise from welding on vehicle bodies caused by:

- fumes
- heat
- electric shocks
- intense light created by the welding process
- sharp edges to metal panels
- trip and slip hazards
- movement of heavy objects.

List of equipment

The following PPE equipment must be worn:

- fume mask – designed to filter out metal fumes
- gloves – heat- and cut-resistant
- welding mask – of a type that is specific to the welding process
- face shield – polycarbonate full-face screen
- ear protection – ear defenders or ear plugs

Grinding metal is also a common cause of injuries arising from:

- sparks
- dust
- projectiles
- noise
- vibration.

FUMES

All paint and electroplating should be removed from the surface of the metal panels to avoid toxic fumes being produced when these are burnt. A fume mask should be worn, or fume extraction used, where there is any risk of fumes being produced during the welding process. Avoid welding close to rubber seals or other combustible materials that are likely to give off noxious fumes when heated.

Fumes created during the welding of metals have been found to cause cancer and there are no guide-

A range of leather gloves used for welding. Left to right: rigger glove; TIG welding glove; and heavy-duty MIG welding glove.

lines at present as to what is a safe level of exposure. An air-fed mask is the best option if carrying out any significant amount of welding.

HEAT

When welding, gloves should always be worn to protect hands and wrists from burns. It is also necessary to protect arms from any burns that may be caused by weld spatter or contact with a hot panel by wearing suitable clothing. **Short-sleeved shirts and short trousers should not be worn whilst welding or working on vehicle bodywork.**

Gloves used for MIG welding should have a high level of heat resistance with long enough cuffs to protect the user's wrists from burns caused by weld spatter that is given off from the process. TIG welding gloves require some heat protection, though need to be flexible enough to operate the torch trigger effectively.

ELECTRIC SHOCKS

Welding machines pose a serious risk of electric shock. Never use in wet, damp or high humidity conditions. Ensure that all cables are kept dry and in good condition. Machines should always be connected to the mains current through an appropriately rated electrical circuit breaker.

It is not advisable to carry out any welding outdoors due to the risks of electrocution and the likelihood of dispersion of shielding gas caused by draughts, as these will result in a poor-quality weld.

Polyurethane-coated gloves should be worn when using chemicals to protect hands from chemical burns and ingestion of harmful liquids through the skin.

INTENSE LIGHT CREATED BY THE WELDING PROCESS

It is important to protect your eyes. The flash of light from the welding process may cause temporary blindness and long-term sight problems if subjected to regular exposure. A suitable face mask is an essential piece of equipment for use during welding. There is a range of varying shades available depending on what type of welding is employed and the amperage used; a higher amperage will produce a brighter flash during welding, requiring a higher level of protection. Metals such as aluminium are highly reflective and a brighter flash will be experienced during the welding process.

There is a wide variety of welding helmets available. For MIG welding, a helmet with a light reactive lens is best, as this enables a clear enough vision of the joint prior to starting the weld. For TIG welding, a helmet with a passive lens is more useful; this lens is always shaded, but has a non-shaded clear area at the bottom of the screen. This clear section of screen enables the welder to see plainly the tightness of the joint before attempting to weld panels together. More detail will be covered on the different masks used for each process in the relevant chapter.

SHARP EDGES TO METAL PANELS

Suitable gloves that offer protection from cuts should always be worn when handling sheet metal. Gloves are available in a wide range of different designs and materials. Some offer more protection against cuts; others more protection from burns or a better grip. The best material that I have found to offer all-round protection and give a reasonable degree of grip is leather. The standard leather rigger glove, which has a safety cuff to protect the wrist, offers better heat protection than thinner welders' gloves. Their disadvantage is in having less control over the torch button. The best of all worlds is to use a rigger glove on the hand holding the panel being welded to protect against burns, and a thinner leather glove on the torch hand to allow for more control over the welding control switch.

CHEMICALS

Chemical-resistant gloves must be worn when working with acids, thinners or paint to protect against the ingestion of chemicals through the skin and to avoid contamination of other surfaces.

A full face screen should be worn to protect eyes and face from splashes and a fume mask should be worn to avoid breathing in harmful fumes.

TRIP AND SLIP HAZARDS

Cables from extension leads, earthing leads and welding torches create trip hazards and should be kept as tidy as possible and put away when not in use.

Any dust produced during the grinding of metals creates a slippery surface when it falls on to a smooth concrete floor. Sweep up dust and filings at regular intervals to reduce the risk of injury from slipping on any debris.

MOVEMENT OF HEAVY OBJECTS

Gas bottles should ideally be stored securely in a mobile trolley to ensure that they cannot be accidentally knocked over and in order to avoid physical injuries that can be caused by lifting or moving such a heavy object. If free-standing, they need to be chained to a wall or a permanent structure. A falling gas bottle presents a risk of crush injury and harm from high-pressure gas that could be released if the valve is damaged or broken off.

Ensure that any vehicle being worked on is adequately supported and that the bodyshell is not able to fall over whilst pushing or pulling against it when fitting parts.

Gas bottles should be secured to a bench or solid structure to avoid the possibility of them being accidentally knocked over.

SPARKS

A serious risk of fire is always present around any welding activity due to the high temperatures attained and the sparks that may be created by the process. Keep a suitable fire extinguisher close to hand, particularly when welding on a complete vehicle that will contain a high proportion of combustible materials.

Oily rags are a common source of a fire starting, as the slightest spark thrown out from the welding or grinding process can cause these to combust. Keep any combustible materials or chemicals in a sealed container to avoid this problem. Welding and grinding should never be carried out near fuel lines or tanks. These should be emptied and purged with Argon gas to avoid any possibility of combustion.

DUST

A suitable mask that is designed to filter out fine dust particles should be worn when using an angle grinder or sanding machine. The adhesives that are used to bond sanding particles to their backing are highly toxic and will be emitted into the atmosphere during the grinding or sanding process. Lead body filler, which is used on older vehicles, is particularly harmful when being sanded or ground, as it is emitted into the atmosphere as fine dust that can be inhaled.

A small fire extinguisher should be kept close to hand when welding, as combustion of materials is an ever present risk, particularly when working on a vehicle body.

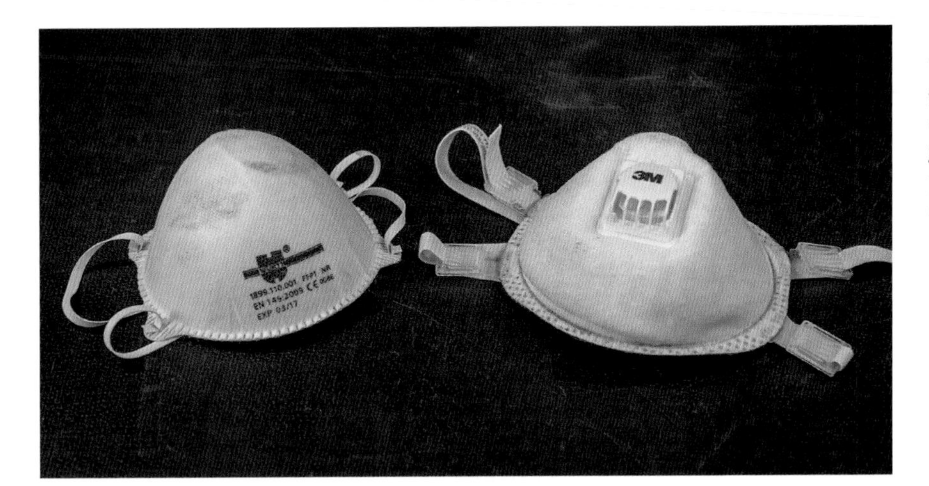

Use a face mask that offers suitable protection from dust particles when grinding (on left), or fumes when welding (on right).

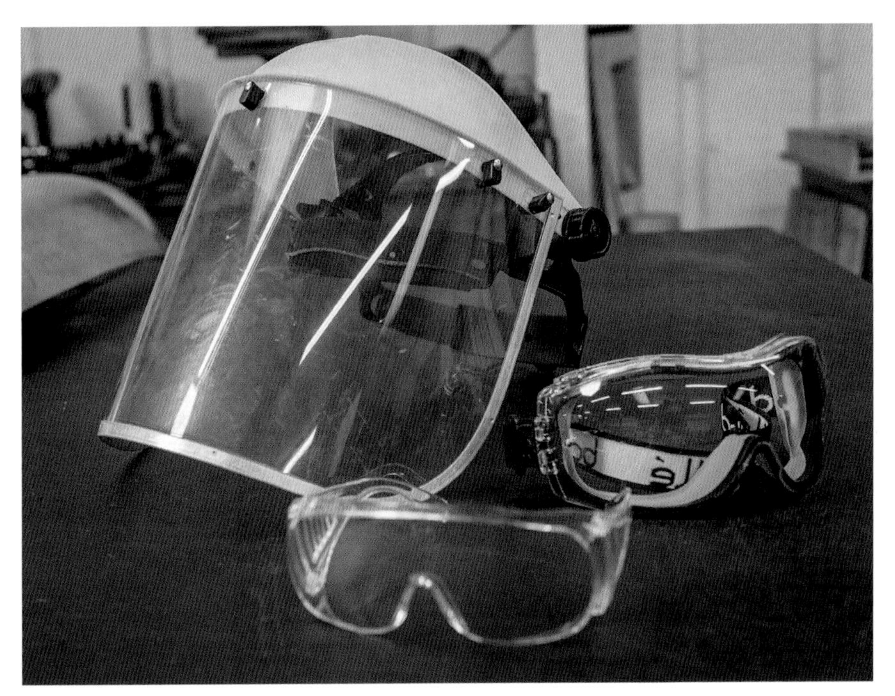

A full-face screen provides the best protection when using a powered angle grinder, which can throw out sharp objects at high speed. Goggles that seal to the face should be used to protect the eyes when general dust is present.

Paints used on older vehicles will contain compounds that are toxic and are particularly harmful when turned into a fine dust that is then readily ingested. Protect yourself and others around you from inhaling these particles by the use of adequate dust extraction and by wearing a suitable filter mask that seals effectively on to your face.

When creating dust, your eyes need to be protected from harmful particles by wearing ski-type safety goggles that seal on to your face.

PROJECTILES

A full-face polycarbonate shield should be used when grinding metals to avoid sparks causing burns and irritation to your eyes and burns or cuts to your face. Fast spinning machines such as angle grinders can throw out particles of metal at very high speed that may cause serious injury to unprotected skin. Protect your whole face from these by using a full-face shield. Be aware of the risk to other people around you, particularly when using an angle grinder.

Ear defenders or ear plugs should be worn when using any machinery that emits a loud or high-pitched noise, such as a hand angle grinder.

A combination of personal protective equipment is needed for work operations that present multiple risks.

NOISE

Angle grinders will cause long-term hearing damage from regular exposure to the high-pitch levels of noise emitted from the machine. Wear ear defenders with a high level of noise protection to avoid any hearing damage. All hearing damage is permanent and will become progressively worse if not protected against. When welding, ear plugs should be worn if there is noise from another source in the workshop.

Chemicals and compounds that are used when cleaning and preparing panels for paint are generally harmful, so all precautions should be taken to avoid breathing in the fumes given off, or allowing the chemical or compound to come into contact with the skin or eyes.

Any waste materials should be disposed of safely. Check with your local council on how to dispose of hazardous waste such as solvents and acids.

Stop and think!

Before you carry out any work operations on a panel or vehicle body, take the time to assess the risks and ensure that you are using the necessary protection to avoid any possible harm.

CHAPTER TWO

EQUIPMENT

When buying tools and equipment it is worthwhile considering the amount of use these are likely to get. Expensive pieces of equipment such as welding machines can be hired at reasonable cost if they are only likely to be used for short periods of time. Second-hand machines will be more affordable to buy, but it is worth getting a qualified electrician to carry out a Portable Appliance Test (PAT) to check for electrical faults before using a second-hand machine.

The most basic or economy welding machines on the market will generally not be adequate. These tend to have a low maximum amperage and although they will have enough power to weld thin sheet metal, they may not operate smoothly on their lower power settings. The sign of a good-quality welder is how smoothly it will work at a low amperage. Mid-range better-quality equipment is usually adequate for sheet-metal work. More expensive equipment is unnecessarily complicated to operate and will not necessarily produce a better a weld.

It is generally worth spending extra money on buying good-quality hand tools, though some tools are over-priced when compared with others of a comparative quality. The price is not always the best indicator of the quality, so carry out some research into what tools other people find work best. Magazine product reviews can be helpful in this regard.

Welding equipment consists of:

- resistance spot-welding machine and range of arms
- MIG welder and torch
- TIG welder – DC (Direct Current) machine for welding mild steel; AC (Alternating Current) machine for welding aluminium alloy
- argon/CO^2 gas bottle for MIG welding
- pure argon gas bottle for TIG welding
- flow meter – for precise control of gas flow to the weld pool.

ELECTRICAL RESISTANCE SPOT-WELDING MACHINES

Spot-welding machines range from small, low-powered, handheld machines to large, high-powered, robotic machines used in car body production. Power is measured in Kilo Volt Amp (KVA). This is the total power output of the machine.

HANDHELD SPOT-WELDING MACHINES

Handheld machines range from 2KVA to 6KVA. For the amateur restorer it is advisable to rent a machine when needed, as the cheaper equipment on the market is usually underpowered and gives no guarantee of achieving a reliable weld. Most small machines will have only one power setting with a variable timer. Basic machines will have no variable controls and rely on the operator to control the time duration of the weld. Older machines tend to be extremely heavy and difficult to handle, whereas more modern equipment is comparatively lightweight. Some machines will have a facility for water-cooling the transformer that improves its efficiency, as a build-up of heat in the machine reduces the conductivity of its electrical components.

SPOT-WELDING ARMS

A range of arms is available for handheld machines. These vary in length and shape to suit the complexity of the panels being joined. These can be homemade from copper bar or adapted to suit the needs of a particular job. It is important to note that there is a power loss over an arm of any significant length due

A handheld spot welder is essential when repairing vehicle bodies, as the majority of panels were originally fitted using this process. A machine with adequate power is expensive to buy, but these can be hired at a reasonable cost.

A wide range of spot-welding arms is available and indispensable for reaching into the complex areas and deep double-curvature shapes that are encountered on vehicle bodies.

to the electrical resistance in the copper bar. Lower-powered machines will only give an effective weld using short arms. Use of longer arms will require a high-KVA output machine to produce a strong weld.

Note that no consumable materials are required for spot welding. The only cost is in replacing worn contact tips that eventually wear down through repeated sharpening, though these will generally last a long time.

ONE-SIDED SPOT WELDERS

One-sided spot-welding machines are commonly used on modern vehicle body repairs. Although they are efficient at welding modern vehicle bodies, they are generally unsuitable for restoration work, as more power than the one-sided machine can offer is required in order to join effectively the thicker metal and lower carbon steel that is used on older vehicles. Higher-carbon steels, as used on later vehicle bodies, require more modern equipment to be used, as these measure the current flow to gauge the effectiveness of the weld, in addition to allowing for slower cooling of the weld area to avoid stress cracks forming around the perimeter of the weld.

MIG WELDING MACHINES AND EQUIPMENT

There are many different makes and models of MIG welding machines on the market, offering you a wide range of control over the welding process. Budget machines will not produce the smoothest of welds due to the cheaper components fitted within them.

POWER OUTPUT CONTROL

All machines will have one main button/dial/switch that can be set in variable or fixed steps to control the power output. This is needed to accommodate the different types and thicknesses of metal to be welded. The power output of different machines will vary. For car body restoration work, a machine with a power output of 80amp is adequate for welding general panel work; however, this will not have enough power for plug and stitch welding effectively even on thin sheet metal. A machine with a 180amp power output is a better buy, as it will also be suitable for welding on a vehicle chassis, as well as carrying out general fabrication work such as making a jig, steel workbench or bespoke tooling.

Some machines will have a spot-weld facility where the maximum power output of the machine is used in conjunction with a timer to facilitate the plug-welding of holes in a panel; this is useful in replicating an original resistance spot weld and is frequently used in restoration work for this purpose.

STITCH WELDING

There may be a stitch-welding control on the machine. This works in conjunction with a timer to give short bursts of high-amperage power to allow welding of thin sheet metal without burning through the joint, or building too much heat into the panel, which would create unwanted distortion.

THERMAL CUT-OUT

MIG machines should be fitted with a thermal cut-out, which will cut power to the transformer if it overheats during operation. There is usually a warning light fitted to the control panel to show when this has been operated. The cut-out will reset at a preset time after the transformer has cooled down sufficiently, in order to avoid damage to the electrical components within the machine.

WIRE FEED

The two diameters of wire used for welding car bodies are: 0.6mm, which is generally used for

A professional quality MIG welder will have a greater range of controls to provide for the different applications of the process.

welding 0.9mm (20swg – Standard Wire Gauge) material or thinner; and 0.8mm, which is generally used for welding 1mm (19swg) or thicker material. Machine settings need to be altered to accommodate the different wire thicknesses. A switch is set to the thickness of the wire being used, which alters the speed of the wire feed.

MIG welding wire can be purchased in various sized rolls, depending on what the machine can accommodate. Larger rolls will require a stronger motor to turn the heavier load placed upon it, so the smaller roll will generally give a smoother weld, particularly on a low-powered machine. The common size welding wire rolls are 5kg and 15kg. All machines have a brake fitted on the wire feed to stop the wire overrunning after the trigger has been released on the torch. This can cause problems on older machines if it is not operating effectively.

The welding wire is fed through rollers that are changeable to correspond with the wire diameter being used. Always make sure that the correct size of roller is fitted to match the wire diameter installed in

RIGHT: ***MIG wire rolls are generally sold in 5kg and 15kg size. Economy machines will generally only accommodate the smaller size roll; professional quality machines will accommodate both sizes of roll.***

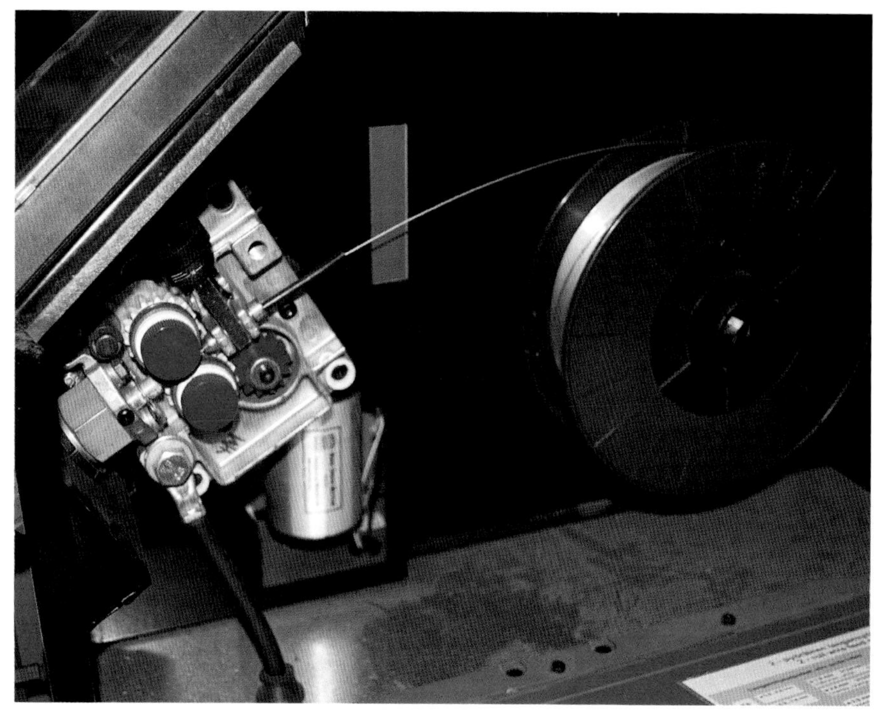

A smooth wire feed is crucial to the MIG process. Economy machines will be found lacking in this area, as the motor used to control the wire feed will not be of a high enough quality to achieve a smooth feed.

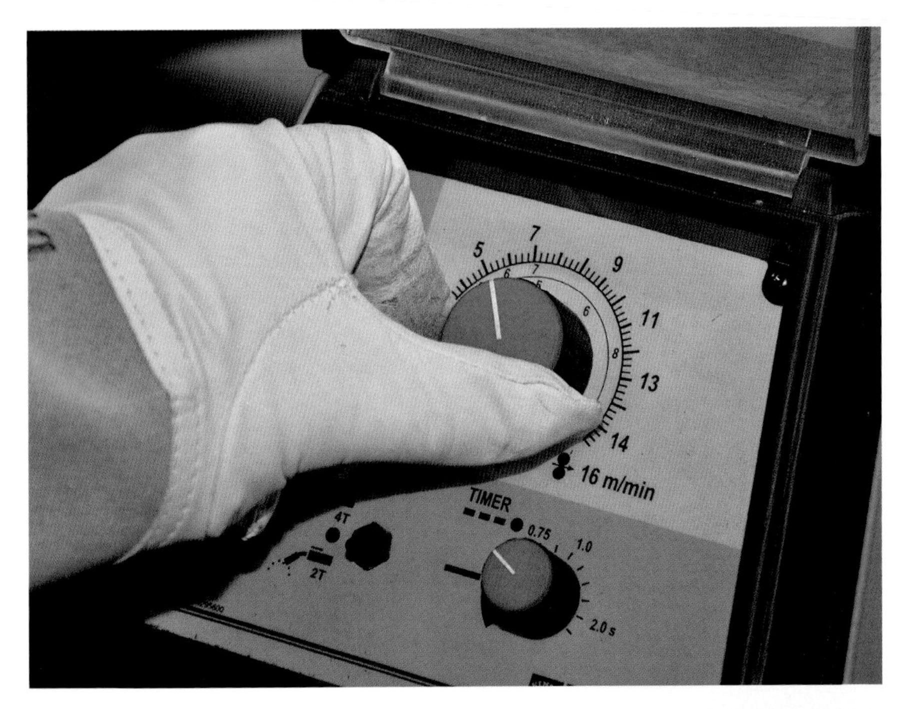

The wire feed is usually controlled independently, though some machines will automatically adjust this to suit the amperage being used.

the machine. The rollers rotate automatically to push the welding wire through the torch lead. The sign of a good-quality machine is a smooth wire feed due to an expensive motor being fitted; economy machines will tend to have poor control over the wire feed due to a budget motor being fitted, which leads to an erratic weld.

Some machines control the wire feed automatically in relation to the power output setting selected. On a machine without automatic wire feed, the speed of the wire is controlled by a separate dial. This needs to be set so that the wire is not pushing the torch away from the material being welded. This happens when the wire speed is too fast. If the wire is burning back into the nozzle of the torch, this means that the wire feed is too slow.

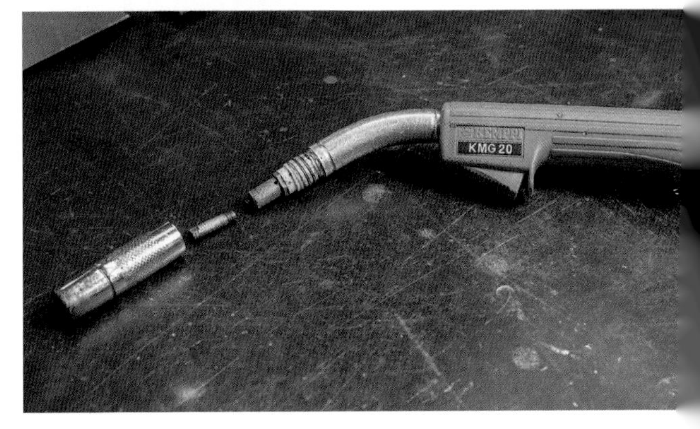

The MIG torch comprises a contact tip that has the relevant bore diameter to suit the diameter of the feed wire being used. This must be kept clear of any weld spatter.

MIG WELDING TORCH

A MIG welding torch consists of:

- a plastic body containing a trigger that operates the electric arc and gas flow simultaneously
- a metal gas shroud or nozzle that contains and directs the gas flow on to the weld area. A separate nozzle is available for plug welding that has two prongs protruding from the end to hold the nozzle the correct distance from the surface of the panel being welded
- a metal contact tip that screws into the main body of the torch. This is of a relevant size to match the diameter of the wire installed on the machine (usually 0.6 or 0.8mm).

The torch trigger control usually has two settings, allowing for:

- 2T sequence – where the gas flow and wire feed start with the trigger being pressed and stop when the trigger is released. If the timer is set, the machine will only operate for the set period of time, with a pause in-between; this facility is used for plug or stitch welding
- 4T sequence – where the gas flow starts when the trigger is pressed and when released the wire feed starts. The trigger is pressed again to stop the wire feed and when released the gas flow is then stopped.

The 2T sequence is most commonly used, as it allows greater control over starting and stopping the weld quickly. The 4T sequence is useful when carrying out a particularly long weld as the trigger can be released, so resting your trigger finger during welding.

Most torches are both replaceable and interchangeable between machines, but will need to have the correct connection to fit the machine's socket.

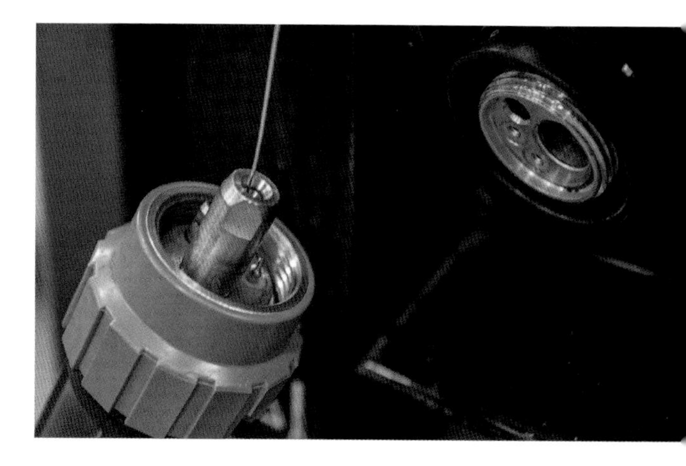

The torch socket on most machines will be of a standard design, though some machines will have their own unique fitting, meaning that the torch is not interchangeable with other manufacturers' products.

TIG WELDING MACHINES AND EQUIPMENT

TIG machines will vary, from those with just a basic amperage control to those having a wide range of controls. They can be further distinguished into those that are only DC output and those that have both AC and DC output. Most metals in common use can be welded successfully with DC, except aluminium and its alloys. This is because when molten, aluminium has a high surface tension that resists the ability for the metal to flow out into a joint. The AC current breaks down this surface tension, allowing the metal to flow more readily when molten, so enabling the welding process.

The simplest DC machines will have an on/off switch and an amperage control. Older machines will be scratch start, which means that the arc is initiated by stroking the surface of the metal with the electrode to allow the current to flow. This can damage

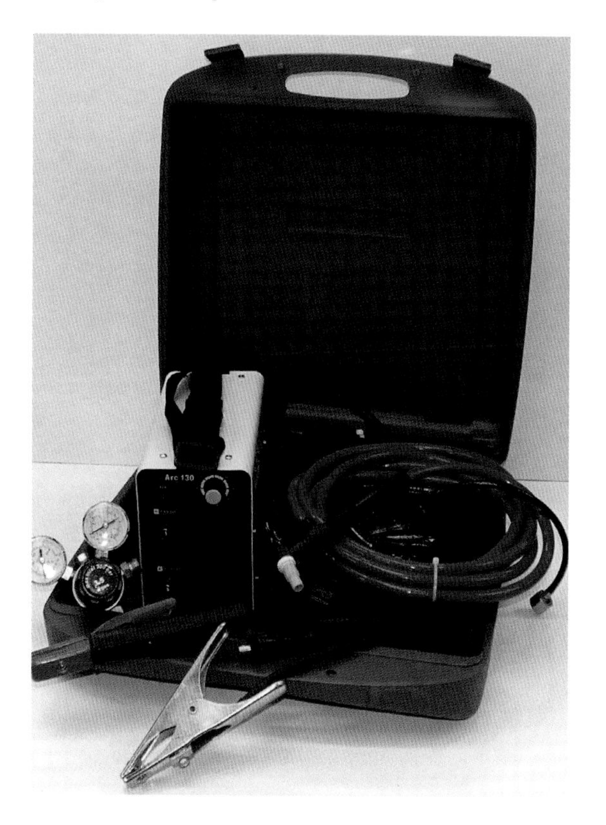

A basic scratch-start DC TIG welding set with a simple amperage control knob is adequate for welding mild steel panels, though some economy machines fail to operate smoothly at the low amperage used.

the point of the electrode, necessitating frequent resharpening.

WELDER CONTROLS

Modern machines have a high-frequency start, similar to the high-frequency spark on a vehicle ignition system. The high-frequency current will jump a gap between the electrode and the surface of the panel, providing a path for the main welding current to follow. This helps to preserve the point of the electrode, which would be damaged by scratch starting.

A foot pedal can be used to control the amperage setting, though this is not recommended for thin sheet metalwork, as the level of control required over the width of the weld is better regulated by the speed of traverse of the torch.

Some machines have a pulse welding facility that is mainly used when welding with filler rod. This provides a timed burst of higher amperage over a regulated period. Rod is melted into the joint during the period of the lower amperage, then withdrawn from the weld before the higher (pulse) amperage kicks in to melt the added metal into the joint. This is necessary to avoid burning the filler rod. This method is usually used for heavier welding applications where additional material is added to create a stronger joint, such as welding brackets to a chassis member.

AC/DC machines will also have controls to vary the frequency and balance of the square wave pattern of the alternating current. Increasing the frequency will narrow the width of the arc, giving a narrower weld. Decreasing the frequency will widen the arc, giving a wider weld. Increasing the square wave balance gives greater penetration of the arc into the weld pool (used when welding thicker materials). Decreasing the square wave frequency provides a greater cleaning effect to the surface of the metal; this is used when a slight surface corrosion is present.

TIG WELDING TORCH

A TIG welding torch consists of:

- a plastic body that contains a tungsten electrode within a metal collet and body
- a ceramic or glass shroud that directs the gas flow over the weld
- a screw-in back cap that tightens the collet around the tungsten and covers the rear of the tungsten.

An AC/DC TIG welder with the full range of controls is essential for welding aluminium alloys. This will also have a high-frequency start facility, making it more convenient for welding mild steel.

R-TECH WELDING

A shorter back cap is available to make the torch more compact, therefore allowing access into tight corners; the tungsten electrode needs to be shortened to accommodate this.

Most torches have a simple on/off switch within the body of the torch. More expensive models have a variable amperage control wheel on the torch. The basic torch is adequate for car body repair work. Flexible head torches are available that allow the head to be positioned at any angle required for ease of use.

GAS LENS

A more recent development is the Pyrex glass gas lens or shroud, which enables the welder to see the arc and weld pool more clearly.

SHIELDING GAS

The shielding gas used in the MIG process is an argon/carbon-dioxide mix in the ratio of around 95:5. The

A standard dual-gauge gas regulator is usually supplied with a MIG or TIG welding machine. This shows the volume of gas contained in the bottle and regulates the pressure that is released. R-TECH WELDING

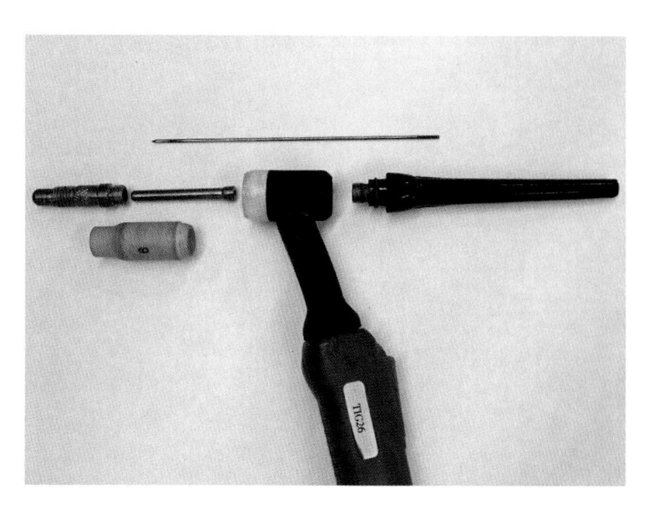

The standard TIG torch contains a metal holder and collet that are related to the size of the tungsten electrode being used and a ceramic shroud to direct the gas flow.

A Pyrex glass gas lens or shroud is a more recent development that enables the welder to see the arc and weld pool more clearly. The tungsten holder and collet need to be of a suitable type to accommodate this.

shielding gas used in the TIG process is pure argon. Most MIG and TIG welding machines are supplied with a dual gauge showing the contents of the gas bottle and the outlet pressure, which can be regulated with a control knob.

FLOW METER

Gas flow is best controlled using a flow meter that accurately limits the flow rate of the gas arriving at the torch. The flow meter is an extra regulator fitted in-line between the bottle pressure regulator and the gas intake on the machine. Never fit a flow meter

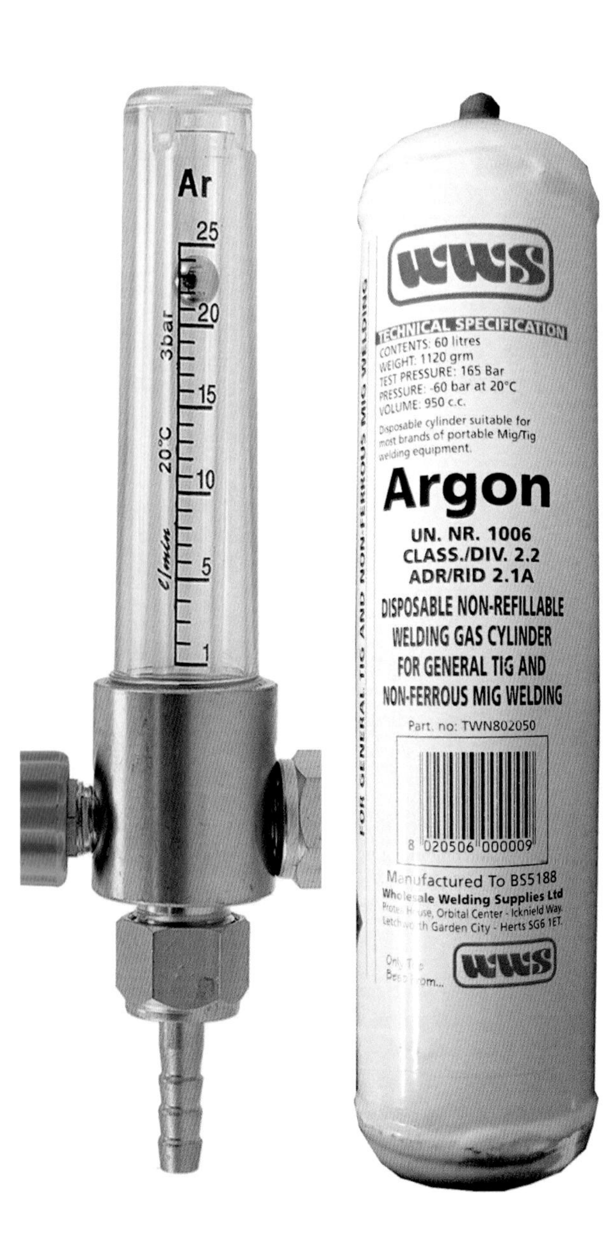

ABOVE LEFT: ***A gas-flow meter, which is graduated in litres per minute, is essential in order to control accurately the quantity of gas delivered to the weld pool.*** R-TECH WELDING

ABOVE RIGHT: ***A small disposable non-refillable gas bottle is an economical alternative to buying or hiring a larger bottle of gas if you are only using a small quantity very occasionally.*** R-TECH WELDING

Maintenance

Welding machines should be regularly checked for:

- a build-up of dust inside the machine casing: welders are often used in close proximity to a grinder, which produces metal dust that can cause short-circuits in electrical components. Regularly remove the casing and blow through the internal components with a compressed air line to remove any dust
- damaged electrical cables: check for loose connections, frayed ends and broken wires or cables
- damaged gas lines: check regularly for leaks in gas hoses and connections; a spray-on leak-detection fluid can be purchased for this purpose. **Always check for kinks in cables or gas hoses, which will impede gas flow, before beginning any welding**
- the power supply to your building will vary, depending on the demands placed on it from neighbouring properties. This can result in a drop in the power output of the welding machine being used; always test the welder on a scrap piece of sheet metal before joining panels to determine its effectiveness. This is particularly important when using a resistance spot welder, as it is often being used at the limit of its power capacity. **What works one day may not work in the same way the next day!**

Regular maintenance of machines and equipment used for welding is important to the welding processes. This will be expanded upon in the various chapters.

directly onto the gas bottle, as the high pressure contained within the bottle will damage the mechanism of the flow meter. The flow rate is set to around 5ltr per minute (10cu ft per hour). Too much gas acts as a coolant, reducing the heat to below that required; it also creates turbulence around the weld pool, causing unwanted inclusions in the weld. Too little gas causes the metal in the weld pool to burn, as the weld pool is not adequately shielded from atmospheric gases.

Gas comes in bottles that can be hired or purchased from a gas supplier and can be refilled when empty. Hiring a bottle reduces the initial cost, as there is a minimal hire fee plus the cost of the gas. Buying a bottle reduces the overall cost over time as no yearly rental is payable, though the cost of the refills may be higher. Various sizes of bottle are available, but be aware that larger bottles are extremely heavy and difficult to handle, so best avoided. There are small disposable bottles available, but these hold a very limited amount of gas and are comparatively expensive. The quality of the gas is occasionally an issue; poor-quality gas, that may contain moisture or impurities, can cause serious problems with the quality of the weld. When connecting a new bottle always test the quality of the gas by carrying out a test weld on a scrap piece of sheet metal before welding on a vehicle panel.

WELDER TROLLEY

A trolley designed for carrying the welding machine and gas bottle is worth investing in, as it enables portability as well as providing safe storage for the gas bottle. Welding consumables and spare parts can also be stored on a trolley to keep everything in one place.

Although most welding machines used for working on thin sheet-metal panels will operate on a standard 13amp fuse, they will usually trip out a standard RCD (Residual Circuit Breaker) commonly fitted to most fuse boxes. A suitable circuit breaker of the correct type is required to avoid this problem. You will need to consult a qualified electrician to determine what type is needed to suit your electrical system.

A welding trolley is useful for storing the welding machine and gas bottle, as well as organizing other welding consumables and equipment in one place.

PRACTICE AND FITTING EQUIPMENT

This consists of:

- welding helmet – suitable for MIG or TIG, depending on which process is used
- torch holder – suitable for holding MIG and TIG torch securely
- welding stool – with adjustable height and tilt mechanism
- body-alignment jig – necessary if carrying out major repairs or panel replacement
- marking tools – indelible marker pen, scriber, blade
- cutting tools – angle grinder, air-powered hacksaw, electric shear
- face shield – for a spot-welding machine
- wire brush – steel-bristled for cleaning mild steel welds; stainless-steel bristled for cleaning aluminium, brass or stainless-steel welds
- clamps – G clamps and locking pliers suitable for clamping joint securely
- small-headed hammer – for levelling tack welds.

WELDING HELMETS

A welding helmet fitted with a light-reactive lens is the best option when MIG welding. These are designed to darken when the arc is struck in order to protect the operator from the glare of the intense light that is given off. They usually have a variable shade setting that can be adjusted in line with the amperage being used; a higher amperage will produce a brighter electric arc and so needs a darker level of shade to protect the user from the glare.

A purpose-designed TIG helmet works best when TIG welding sheet metal, as it allows a clear vision through the bottom of the screen when it is partially open. This is necessary to ensure that the joint is seen to be tight during butt welding. Use a no.9 glass shade on the helmet for the amperages used for car-body repair work. Helmets are usually supplied with a no.10 shade; this is fine for heavier applications, but too dark for the amperage used for welding thin sheet metal. **Seeing the weld pool clearly is critical to achieving a good weld.**

The most useful welding helmet for the MIG process is one with a light-reactive lens. This should have a variable control to set the darkness of the shade depending on the amperage being used.

TORCH HOLDER

Use a torch holder to avoid the torch dropping to the floor, particularly as the ceramic or glass shroud fitted to the torch is fragile and easily cracked. A torch holder with a magnetic base is particularly useful, as it can be stuck to a steel bench or a body panel when working on a vehicle.

WELDING STOOL

An adjustable height swivel stool is a must to enable the welder to sit in a comfortable position whilst welding, either working on the bench or on a vehicle body. The better-quality stools will also have an adjustment for the angle of the seat, making it more comfortable.

A helmet specifically designed for TIG welding allows a clear vision through the bottom section of the mask. This is more critical when butt welding in order to ensure that the joint being welded is seen to be close-fitting.

JIGS

Body-alignment jig

Some form of body jig is necessary if carrying out any major panel replacement in order to ensure that the body alignment is maintained, particularly on a vehicle without a substantial chassis to support

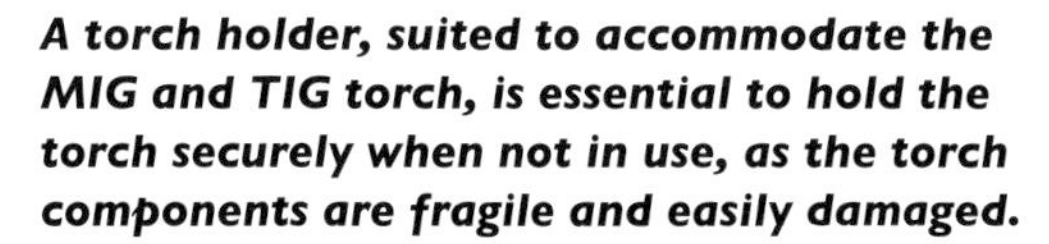

A torch holder, suited to accommodate the MIG and TIG torch, is essential to hold the torch securely when not in use, as the torch components are fragile and easily damaged.

A comfortable welding stool with an adjustable height and tilt mechanism is essential for achieving a satisfactorily smooth weld when working at a bench or on a vehicle body.

A jig for fitting the bodyshell during welding is essential if carrying out any major repairs or replacing structurally important panels such as sills.

A rollover frame can be fitted to a body alignment jig to allow the positioning of the body at a convenient angle for carrying out repairs.
KEITH PELLEN

the bodywork. The vehicle's suspension and transmission mounting points should be incorporated. Measurements for these points are sometimes available from the original workshop manual for the vehicle. Heavy-duty castors can be fitted to the jig to allow it to be moved around the workshop; these should be fitted with a braking mechanism for safety.

Rollover jigs

A rollover jig, or rollover frame, fitted to a body-alignment jig is particularly useful when welding on vehicle bodies. This allows you to position the bodyshell at the most convenient angle to carry out welding and repairs in the most comfortable and safest position. A secure locking mechanism must be fitted to ensure that the body cannot move during repair work.

MARKING TOOLS

Indelible ink marker pens are useful for the initial marking of panels to indicate where any excess waste material needs to be removed. A clearer view of a scribed line on sheet metal is given if a pen line is initially marked on the panel; the scribed line is then highlighted, making it easier to see.

Scriber

A hardened-steel scriber with a long straight and hooked end is used for marking out on sheet metal to give a precise cut line on a joint. These are easily lost because they are small and dull in colour. An idea that I use is either to dip the tool in a bright paint, or cover it with a brightly coloured tape to make it more visible.

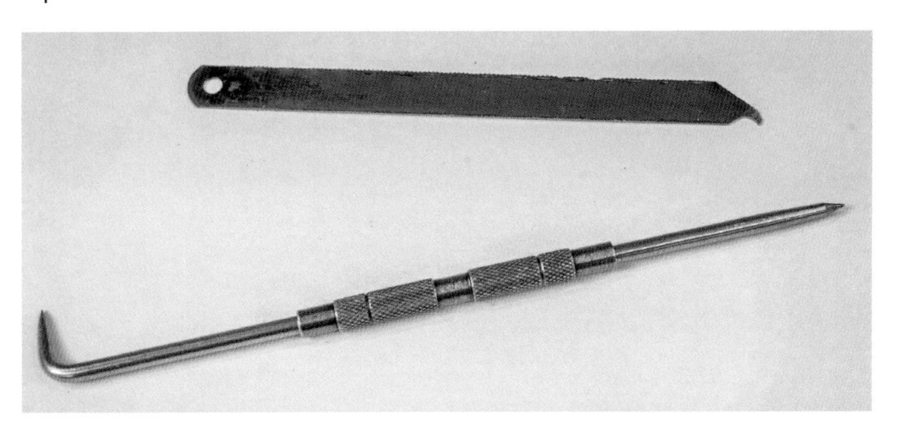

A hardened-steel scriber is used for general marking out on sheet metal. A straight-edged blade is used for marking from edges of repair panels to establish a cut line.

A straight-edged blade is indispensable for marking accurately from the edge of a panel to give a clear-cut line for welding in repair sections. An old hacksaw blade can be used for this purpose.

PORTABLE POWER TOOLS

Various electrically and air-powered portable power tools are available for cutting sheet metal. These all have their advantages and disadvantages.

Electrically powered tools

Electrically powered tools may be mains powered or battery powered. Although battery-powered tools are becoming more efficient, the disadvantages compared to corded tools are: a comparative lack of power; limited use time; and heavier weight due to the addition of the battery; this will cause fatigue over time and is more likely to lead to repetitive-strain injuries.

Pneumatic-powered tools

Pneumatic-powered tools (that is, powered by air) are commonly used on vehicle bodywork, as they offer several advantages over electrically powered tools:

- they are lighter, so reducing fatigue in operation
- they won't overheat, as the compressed air that powers the motor keeps it cool
- they have a fast start and stop time, making them more efficient and safer to use
- they tend to be more durable as they have less components to break or wear out.

Pneumatic tools will require an adequate air supply to run them effectively. The power output of a compressor is measured in Cubic Feet per Metre (CFM). The minimum power required for operating the handheld tools used in restoration work is 8CFM. An air compressor is a useful addition to any workshop, as it can also be used for spray-painting panels and other components.

Air-powered hacksaw

An air-powered hacksaw is one of the most useful tools for cutting thin sheet, as the high speed of the blade avoids the tool jumping on the panel, which will be experienced when using an electrically powered jigsaw because this operates at a much lower speed. The high operating speed of the air-powered saw will cause the blades to blunt relatively quickly, so these will need changing regularly.

An air-powered hacksaw is commonly used for removing panels from a vehicle body, as the high speed of the machine means that it cuts cleanly without distorting thin sheet metal.

Reciprocating saws

A reciprocating saw is useful where precision is needed in cutting a straight line and for getting into tight corners that cannot be reached by other tools. These are available either air- or electrically operated and can be fitted with a range of various-shaped blades to suit the joint being cut.

Grinding machines

Grinding machines, which may be air- or electrically powered, are useful tools for cutting out areas of corroded metalwork and for finishing welded joints. A range of machines in various sizes are available that can be fitted with thin discs for cutting sheet metal, hard wheels for grinding and card or fleece discs for sanding and finishing.

Reciprocating saws are useful for cutting out certain areas of panel work that are difficult to access. They are not so good for general cutting work, as they are slow and the blades are comparatively expensive to replace.

Handheld grinding machines, which may be air- or electrically powered, are a useful tool for cutting out areas of corroded metalwork, as well as for finishing welded joints

Electric shears

An electrically or air-powered slot-cutting shear will cut through the middle of a large panel without causing distortion. This machine is particularly useful for removing corroded panels.

An electrically or air-powered side-cutting shear will cut thin sheet metal quickly and is good for removing the majority of the waste material on the edge of a panel before cutting accurately to a scribed line using hand tin snips.

An electrically or air-powered slot-cutting shear will cut through the middle of a large panel without causing distortion. This machine is particularly useful for removing corroded panels.

HAND TOOLS

Sheet-metal shears

Hand-operated shears are a necessity for cutting the edges of sheet-metal panels accurately, particularly when cutting to a curved line. These shears are available as:

- right-hand snip: used for cutting from right to left on a curved line

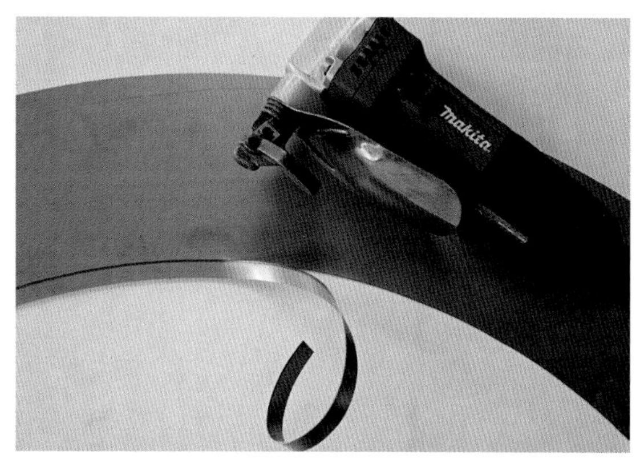

An electrically or air-powered side-cutting shear will remove metal quickly and is good for removing the majority of the waste material on the edge of a panel.

Hand tin snips are best used for cutting accurately to a marked line. These are available with blades for cutting right hand and left hand on a curved line, as well as straight blades for cutting on a straight line.

- left-hand snip: used for cutting from left to right on a curved line
- straight snip: used for cutting on a straight or very shallow curved line
- these need to be kept sharp and require a degree of practice in their use to cut accurately to a scribed line without distorting the panel.

Drilling

Drilling circular holes in thin sheet metal is best done using initially a small twist drill, 1.5mm (1/16in) to 3mm (1/8in) in diameter and then progressing on to a cone or stepped drill to increase the size of the hole. Larger diameter twist drills will not cut a satisfactory round hole in sheet metal due to the geometry of the drill flutes clashing with the thin sheet.

Punching

A hand-operated hole punch will cut small diameter holes cleanly and is particularly useful in making holes to create plug welds. These have a range of punch diameters up to 6.4mm (1/4in) in diameter.

Large diameter holes are best cut out using a metal 'chassis punch'. The commonly available punches are those used by electricians to cut holes in metal trunking. These work by screwing the punch

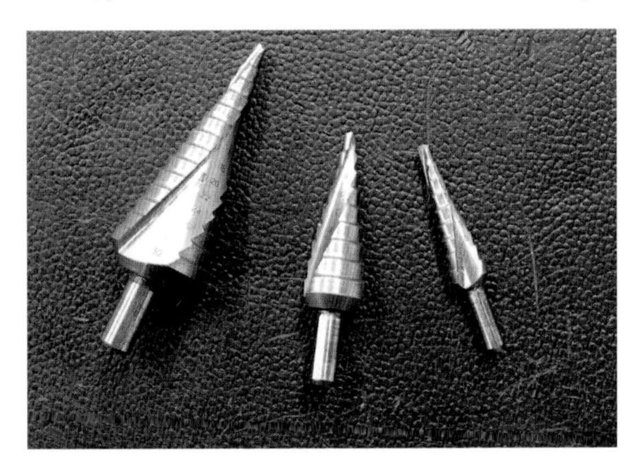

A range of hole-cutting tools is available. The most efficient drills at producing a clean round hole on thin sheet metal are stepped or cone drills.

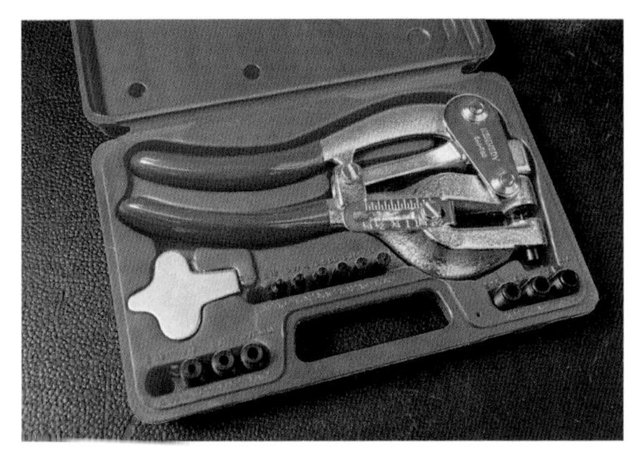

A hand-operated hole punch will cut small diameter holes cleanly and is particularly useful in making holes to create plug welds.

Larger holes are best cut out using a chassis punch. These are available in a wide range of metric and imperial sizes.

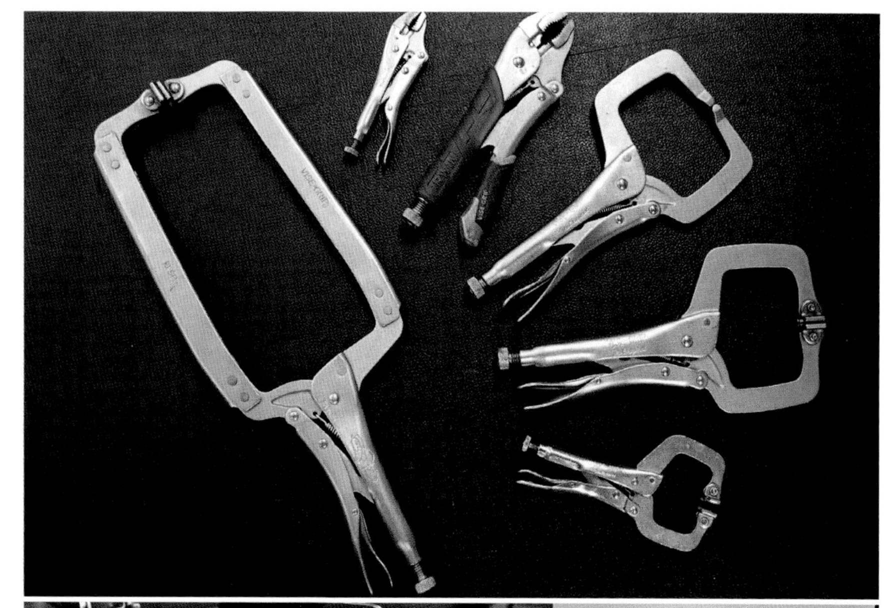

A range of clamps is needed to hold panels securely in place when marking out joints and for holding in place during tack welding.

A clamp that holds the locking plier to the bench is useful for holding panels when welding or cutting, as the panel can be held in a convenient position.

into a corresponding die. A pilot hole, the diameter of the punch screw, is initially drilled into the sheet to provide a central location for the punch and die and accommodate the centre screw. The screw should be well greased to avoid damage to the thread, which is placed under severe strain in use.

Clamps

There is a huge range of clamps and locking pliers available of various sizes and shapes specifically designed for holding car body panels during assembly. It is well worth buying good-quality clamps and locking pliers in particular, as less expensive ones create problems, causing misalignment of panels due to poor clamping pressure and can be generally difficult to use.

Locking plier clamp

A clamp that holds the locking plier to the bench allows the panel to be held in a more convenient position for welding or cutting when trimming edges.

Files

A range of hand files is needed to work on sheet metal. For mild steel, a smooth-cut, half- round file is best for general purpose use. All files should be stored on a wooden shelf, or wrapped in cloth to avoid damage to the teeth from other hard metal tools.

Body file

A body file that has an adjustable frame allowing the blade to be bent into a concave or convex curve is useful for checking the smoothness of a panel's surface.

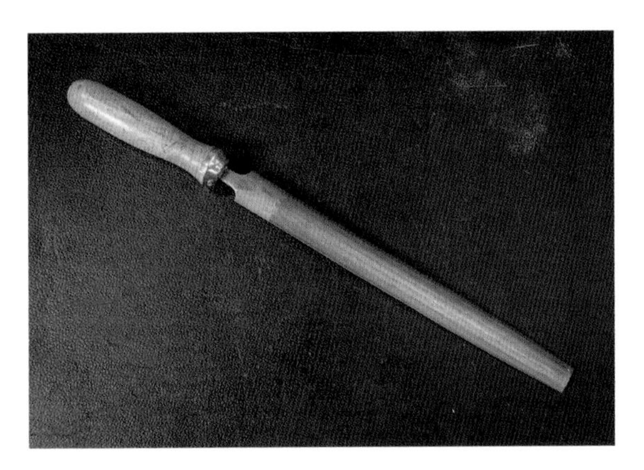

A smooth-cut, half-round file is best for general purpose use in preparing the edges of the panels for welding.

A file card or brass wire brush is used to clean the file teeth. This should be carried out regularly to clear teeth of metal filings.

Soft metals such as aluminium alloy require a coarse file to avoid the teeth of the file clogging up with the soft metal particles that are removed. The coarsest file available is known as a dreadnought. This has single lines of curved teeth. The file should be cleared regularly of any metal filings using a file card or brass wire brush. A dreadnought blade can be purchased that fits into a handle. This is able to be curved in either direction to file a convex or concave surface or edge (commonly known as a body file).

A coarse or dreadnought file is used for filing a soft metal such as aluminium alloy. A large blade is available that fits to a body file allowing the blade to be curved slightly in either direction.

FINISHING EQUIPMENT

This consists of:

- small-headed hammer
- planishing hammer
- flipper/body spoon
- dollies
- angle grinder
- power sander
- electric hand drill and drill bits
- body file.

HAMMERS

A range of different hammers with various-shaped heads and weights is needed to carry out work effectively on double-curvature sheet-metal panels. A small-headed flat-faced hammer is needed to work initially on the welded joint. The small head ensures accuracy. It is generally better to work the metal slowly, so avoid using too heavy a hammer for this purpose, as it is likely to create distortion by over-stretching the metal.

The flat-faced or planishing hammer is the most commonly used tool for final finishing. The larger head helps in smoothing the panel without over-stretching the metal.

A small-headed hammer is used initially to flatten and harden welded joints. The small head ensures accuracy in hammering directly on the weld.

A large flat-faced hammer is used for smoothing welded joints. The hammer's face is ground and polished to create a very slight convex shape to its surface.

FLIPPER OR BODY SPOON

A flipper or body spoon is used particularly on aluminium alloy panels to smooth the surface. This tool offers a larger surface area from the face of the tool than a planishing hammer and avoids marking the surface of the panel and overstretching the metal. A flipper is either straight along its length, or curved to suit the various shapes of panels being worked on. These were traditionally handmade from an old car leaf spring or large file that would be heated and bent into the desired shape and ground smooth, but there is a large range of manufactured tools now available.

DOLLIES

Forged-steel dollies need to be shaped to fit the curvature of the panel being worked on. The most useful profile shape of dolly is one with a hand grip that allows it to be held firmly. It is best to have several dollies of this type that can be ground to different profiles using a hand angle grinder to suit the shape of the panel.

A flipper or body spoon is used particularly on aluminium alloy panels to smooth the surface. Straight- and curved-faced tools are needed to suit the various shapes of panels.

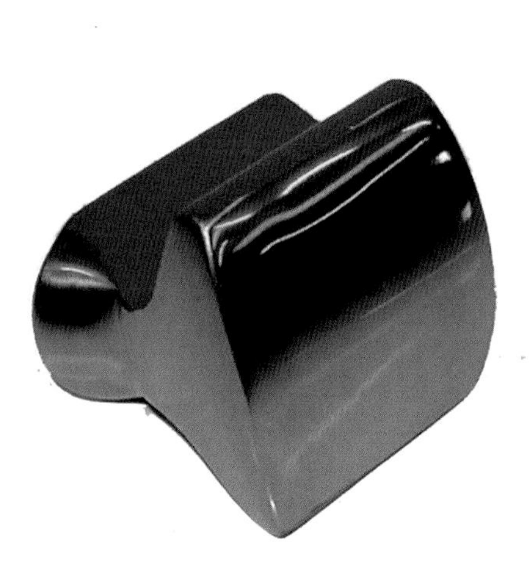

Handheld dollies are indispensable for working on sheet-metal panels. This is the best form of dolly for general use and is ground to suit the shape of the panel being worked on.

SANDERS

To achieve the required finish for painting and so on, small belt and drum sanders are useful tools, particularly for finishing work on the more complicated detailed shapes that are commonly found on vehicle bodywork.

A random orbital sander will create a smooth finish to a surface, particularly on a large, flat or shallow double-curvature panel. This can be fitted with a range of sanding discs, as well as polishing mops for cleaning paintwork.

Small belt and drum sanders are useful tools, particularly for finishing work on more complicated detailed shapes.

A random orbital sander's action is particularly effective at creating a smooth finish to a surface, especially on large, flat or shallow double-curvature panels.

STORAGE

All tools and equipment should be stored safely to avoid loss and damage. A tool hanging rack or clipboard helps in finding tools quickly. It is good to develop a habit of returning any tools to the rack or board immediately after you have used them so that your workplace is kept clear of tools and they are not lost or damaged.

Measuring and marking-out tools, which are more prone to damage that will impair their accuracy, need to be stored more securely, so are best kept in a tool cupboard or drawer within a purpose-made cut-out board.

BELOW: ***Hand tools that are used regularly can be stored using a peg or clipboard so that they are close to hand and easily accessible.***

BELOW OPPOSITE: ***Measuring and marking-out tools, which are more prone to damage, need to be stored more securely and are best kept in a tool cupboard or drawer within a purpose-made cut-out board.***

Tool maintenance

All hand tools should be regularly maintained to ensure that they are working effectively:

- cutting and marking out tools need to be kept sharp
- clamps and locking pliers should be greased on threaded and moving parts
- measuring tools should be checked for accuracy
- hammer and dolly faces need to be kept smooth
- power tools should be checked for loose parts and damage to cables and plugs.

CHAPTER THREE

GENERAL PRINCIPLES OF WELDING

Welding is simply the melting of two separate pieces of metal at their junction to allow the materials to flow together to form a single piece of metal when cooled. Enough heat must be created at the joint to enable the metal to flow effectively so that this process can take place. At the same time, the molten metal must be shielded from the surrounding atmospheric gases. This is to avoid contamination of the weld pool from nitrogen and burning of the metal that requires oxygen. Oxygen must be present for any combustion to take place. Different metals will behave differently when molten, depending on their degree of surface tension when in the liquid state.

The difficulty in learning to weld, which is often experienced by the beginner, is partly due to not understanding fully how metals behave and the limitations of the process being used to create a welded joint.

BEHAVIOUR OF METALS

Mild steel, which is most commonly used for vehicle body panels, readily flows when molten. Aluminium and its alloys are seen as notoriously difficult materials to weld, which is mainly due to their high surface tension when liquid. This means that the metal does not flow readily when molten due to an extremely hard oxide layer forming on its surface when in contact with atmospheric oxygen. This problem is overcome by the use of a flux when oxyacetylene welding, or the use of AC when TIG welding. The flux and AC current break down the surface oxide, allowing the metal to flow more readily when in the liquid state.

SKILL LEVEL

While the processes of MIG and electrical resistance spot welding require very little skill in their operation, TIG welding requires a comparatively high level of skill in developing the techniques needed to make a successful weld, particularly when welding aluminium and its alloys. This means spending significant time in practice to attain a good level of control over directing the aim of the welding torch and in improving physical reactions to control the speed of traverse. A clear vision of the weld pool is also critical to the process. It takes practice to discern how close you need to be to the weld pool to see it clearly. However, there is no reason why anyone should not be able to weld cleanly and effectively once they have learnt and practised the necessary techniques.

TYPES OF WELD JOINT

The two common joints used on vehicle body panels are lap and butt joints.

LAP JOINT

The lap joint is where the edges of the panels are overlapped, either sitting flat on top of each other or meeting squarely as folded flanged edges. These are commonly resistance spot-welded. This is the most common form of joint found on mass produced vehicle bodies as it facilitates relatively easy and accurate assembly of panels.

The joggled or stepped joint is a form of overlapped joint that is commonly used on thin sheet metal, particularly in restoration work, as it stiffens the panel on the joint and requires less precision in assembly than other forms of joint.

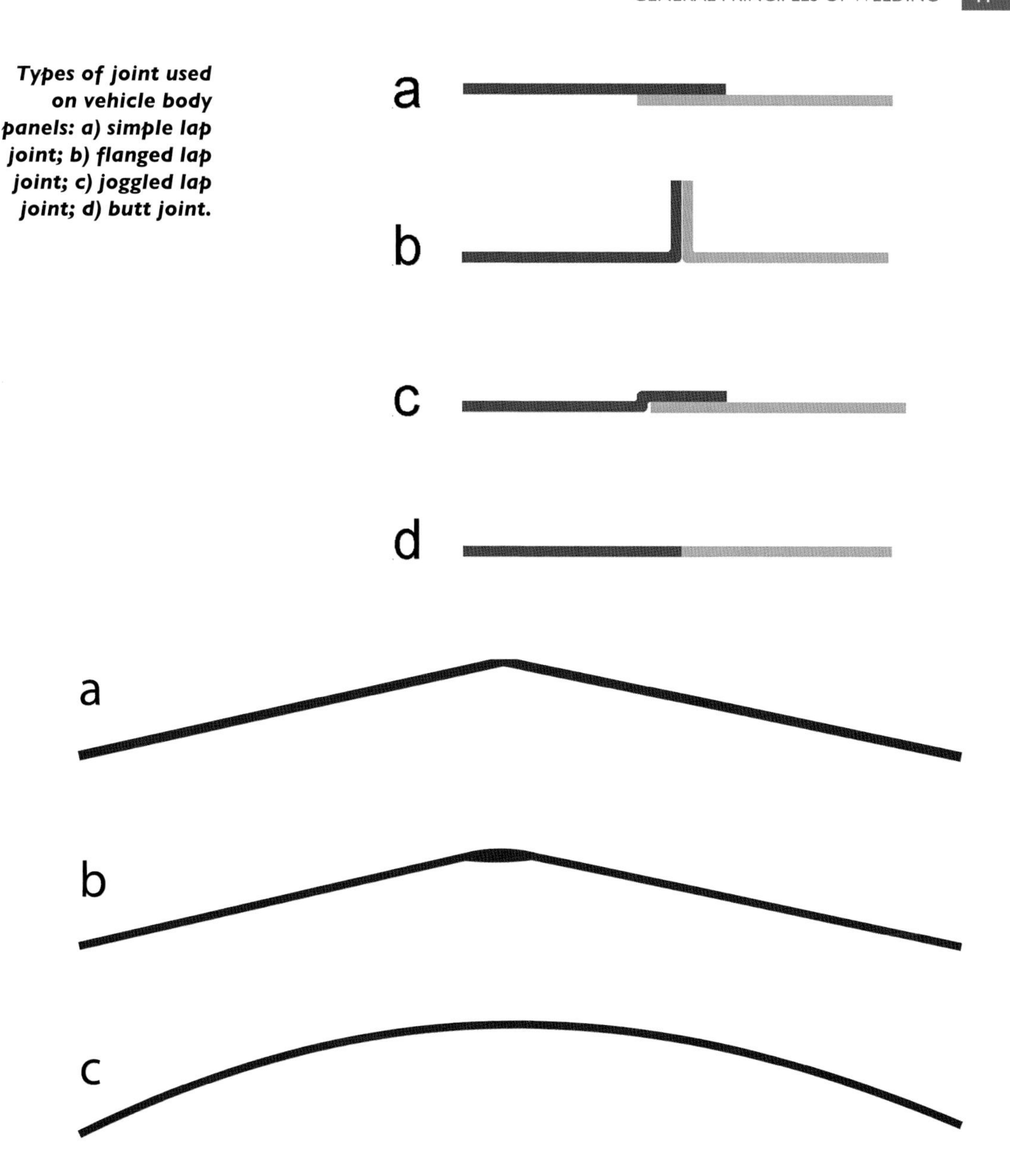

Types of joint used on vehicle body panels: a) simple lap joint; b) flanged lap joint; c) joggled lap joint; d) butt joint.

Effects of thickness of weld when placed under stress: a) if the metal is thinner or softer on the weld the panel will bend, creating a crease on the joint; b) if the metal is thicker or harder on the weld the panel will bend, creating a crease around the joint; c) if the metal is the same thickness and hardness on the joint the panel will bend without creasing.

BUTT JOINT

A butt joint is where the panels meet edge to edge with no overlap. This involves using oxyacetylene or TIG electric-arc welding to fuse the edges of the panels together. This type of joint is most commonly used in repairing original panels by welding in patches or part panels. It requires a high degree of precision in cutting the edges of the panels so that they meet accurately. It is essential when butt welding parts together that the thickness of the metal is not increased or decreased, as this will seriously affect the smoothness and regularity of the contour you are aiming to produce.

The use of a filler rod is to be avoided, as it will increase the thickness at the joint and will be detrimental to the finish of the panel. A filler rod is only used where extra metal is required, such as on the edge of a lap joint. On a butt joint the two separate pieces of metal are simply fused together using the TIG process.

If the metal is thinner or softer on the joint, this will cause the panel to bend at this point if placed under any stress. If the metal is thicker or harder on the joint, the panel will bend either side of the joint if placed under any stress. Most vehicle body panels are placed under stress when the vehicle is in motion, so if the metal is not kept to the same thickness and hardness as the rest of the panel the weld line will eventually show as a kink or fold in the panel's surface.

WELDING PROCESSES USED IN VEHICLE BODY RESTORATION

A concentrated heat source of a high enough temperature to melt the metal being welded is needed to achieve a reliable join between separate pieces of metal. The heat can be created by combustible gases, an electric arc, or by using electrical resistance. Oxyacetylene gas has been traditionally used in the making and repair of vehicle body panels, as it can readily achieve temperatures high enough to melt steel. MIG and TIG gas-shielded electric-arc welding has generally superseded the use of oxyacetylene because of its many advantages: it is safer in use; it creates less distortion due to a more concentrated heat source; and it negates the need for flux when welding aluminium alloys, as the action of the alternating electrical current breaks down the surface oxide of the metal.

The majority of joints on vehicle bodies were originally spot-welded, as this is the most efficient method of joining panels and creates the least distortion.

RESISTANCE SPOT WELDING

The majority of vehicle bodies were originally assembled using electrical resistance spot welding due to its efficiency, ease of application and the relatively low heat produced in operation. Other methods of welding such as oxyacetylene, MIG and TIG create a wider spread of heat into the panels, causing distortion as the metal expands and contracts.

Resistance spot welding is generally only employed on lap joints, as very specialist equipment is needed to create a resistance-welded butt joint. A stitch welder was used on mass-produced body panels to join large pressed double-curvature panels using a very small overlap on the joint. This is an electrical resistance welder that creates a continuous weld by using rollers, through which a high-amperage current is passed and fed over the joint, compressing it and fusing the two panels together. The resulting bead of weld is then ground level on the outer face of the panel, but left as a visible bead of weld on the reverse surface of the panel, which is not seen on the finished body.

Original manufacturer's stitch-weld. The front and rear half of a wing panel are formed as two separate pressings, then overlapped and welded to form one panel.

METAL INERT GAS

MIG welding is most commonly used for repairs to vehicle bodywork due to its ease of use. A filler

MIG welding is commonly used for repairs due to its comparative simplicity in operation. However, it can create significant distortion due to a build-up of excessive heat during the process.

TIG welding is particularly useful for butt welding in repair patches or larger panels as the joint can be planished to a smooth finish, leaving an invisible repair following painting.

wire, which acts as an electrode, carries the electrical current to the weld pool. However, problems can be created due to distortion caused by building excessive heat into the panels through the addition of molten metal (the filler wire).

TUNGSTEN INERT GAS

TIG welding, which has directly replaced oxyacetylene, is particularly useful in the restoration of vehicle body work when butt welding a repair patch into panels, as a seamless joint can be achieved with minimal distortion created. If no filler rod is used, the joint is easily planished smooth following welding. This can be DC or AC, depending on the material being welded.

SHIELDING GASES

MIG and TIG methods both use a shielding gas to avoid the inclusion of atmospheric gases into the weld pool, which would weaken the welded joint due to the aerated nature of the gas inclusions. The gas used for the MIG process on mild steel is an argon/CO^2 mixture in the ratio of 95:5%. This provides the best result in the finish and structure of the finished weld. Pure argon gas offers the best results for the MIG process when welding aluminium alloy and for the TIG process on all metals. The shielding gas also provides a plasma (ion neutral medium) for the electric arc to pass from the electrode to the material being welded. Argon is a noble gas, which means that it does not react with other elements. Both gases are heavier than air, odourless and invisible, which makes them dangerous in that they are not seen or smelt, but are asphyxiants that will displace the atmospheric oxygen. For this reason, it is important to be extra vigilant in ensuring there are no gas leaks from the apparatus, particularly when working in a confined space. Welding will create a build-up of harmful gas over time, so should not be carried out in a confined space without the air being replaced in some way.

Most metals used on vehicle bodies can be welded using all the above methods with varying degrees of difficulty dependent upon the particular metal's characteristics.

PREPARING PANELS FOR WELDING

It is important to ensure that all panel joints are a good fit before welding the panels in place, as any gaps in the joints will create unnecessary distortion as the added metal cools and shrinks. This can also pull panels out of alignment, creating problems with the squareness of the welded structure. All paint, electroplating and oxide should be removed, as these will adversely affect the strength of the welded joint by contaminating the weld pool. A weld-through, zinc-rich primer can be applied to the mating surfaces of any panels being resistance spot-welded in order to protect the metal on the joint from future corrosion. Be aware that any gases produced from the primer burning during welding are toxic, so a suitable fume mask must be worn to avoid inhaling the toxic fumes that are given off.

TACKING AND RUNNING-IN THE WELD

All welded joints are firstly tack-welded to hold the panel in place before the final process of running-in the joint. In the MIG and TIG process, the tacks are small points of weld equally spaced along the joint to hold the panels level and secure them in place prior to making the final weld. The running-in process involves running a continuous line of weld along the tacked joint. In the resistance spot weld process, the tacks are single spot welds placed at the end and centre of the joint, initially to hold the panel in place. Further spot welds are placed in-between these initial welds until the panels are securely joined.

If the panels are not tack-welded sufficiently securely, the expansion in the metal, caused by the heat produced during the running-in process, can move the joints out of alignment.

MINIMIZING DISTORTION

Distortion from the welding process is caused by heat shrinkage on and around the joint. As soon as any heat is applied to a metal it expands and during welding is then joined whilst in this expanded state. As the metal cools the welded joint and surrounding panels shrink, causing distortion as the panels are now held in extreme tension. The thickness of the panel is also now greater on the welded area.

To minimize the distortion, it is best to limit the amount of heat spreading out into the surrounding panel. This is achieved by welding the joint as quickly as possible and keeping the weld as narrow as possible. The mistake that the beginner often makes is to use less amperage in order to avoid blowing holes; this means that a longer time is needed to melt the metal, which results in a wider spread of heat that produces a greater degree of distortion as the metal cools and shrinks. The metal also won't flow so readily using a lower amperage, making it more difficult to create the initial tack welds. It is best to practise welding on scrap material, working as quickly as possible so as to become adept with working at speed using a high amperage.

The consistency of the width of the weld is crucial in order to avoid building excessive distortion into the surrounding panel. The area of blue heat spread is a good indicator of this.

The thickness of the metal on the welded joint increases due to the expansion and contraction caused by the heat induced during the process. This creates distortion as the surrounding panels are pulled into tension.

When joining panels of different thicknesses, bias the welding arc towards the thicker metal so that both panels melt at the same time.

A consistent width of weld will cause less distortion than an inconsistent width. The blue area around the weld, which is visible following welding, is a good indicator of the consistency achieved. Aim to keep this blue area of heat spread as even in width as possible over the entire length of the weld.

The distortion that occurs following any welding is caused by the metal shrinking following the welding process and thickening on the welded joint. This is resolved by hammering the weld back to its original thickness, which relieves the tension across the panel and so alleviates the distortion.

WELDING PANELS OF DIFFERENT THICKNESSES

When welding panels of different thicknesses, which is often experienced when fitting repair panels or patches, it is necessary to avoid the thinner metal melting before the thicker metal. The aim is to melt the edges of both panels at the same time. In order to achieve this on a butt joint, the thicker material will require more heat input. To build more heat into the thicker material, bias the arc or heat source towards the thicker panel as you are welding the joint.

Resistance spot welding panels of different thicknesses on a lap joint creates no real problems, as both panels will reach the same temperature at the fused joint.

TESTING WELDS

The strength of welds on all the processes outlined should be checked to ensure that they won't fail under normal stress loads.

To check if a weld is strong enough you need to tear a welded joint forcibly apart (on a test piece using spare metal – not the actual panels to be welded). Sample welds of all methods should be tested to destruction to ensure that the welded joint is strong enough to withstand the stresses it will be placed under when on the vehicle.

TESTING A RESISTANCE SPOT WELD ON A LAP JOINT

Clamp one panel firmly in a vice and use locking pliers clamped to the second panel to peel the joint apart. It should require considerable force to break the weld and a hole should be left in one panel and the metal from the hole left firmly attached to the other panel to show that the weld has been effective.

TESTING A MIG PLUG WELD

Use the same method to test a MIG plug weld as used to test the resistance spot weld (*see* overleaf).

TESTING A BUTT WELD

Clamp one panel of the test piece in a vice just below the welded joint, then using locking pliers attached

Resistance spot welds and MIG plug welds are tested by physically pulling the joint apart. A successful weld will result in the metal from one panel being left attached to the other panel.

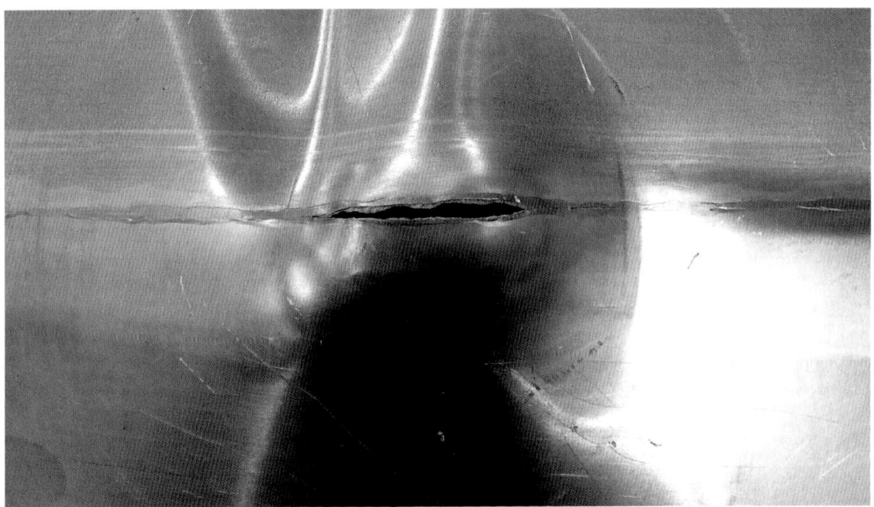

A MIG or TIG butt weld is tested by hammering on the joint with a ball-headed hammer into a dish-shaped wooden block or sandbag. If the joint fails using reasonable force, the weld was not successful.

The same test using considerable force shows that this weld is secure, and the process was successful in producing a joint of equal strength as the surrounding panel.

It is worth establishing an area in the workshop where you can practise the various techniques and processes. Butt welding, in particular, demands time spent in practice to become skilled in carrying this out effectively.

to the other half of the panel above the welded joint, bend the panel, moving it backwards and forwards repeatedly. There should be no fracture on the welded joint.Any fracture on the joint shows that the weld has not been effective. It should require several bends with the metal visibly tearing randomly across the panel and leaving a jagged edge to show that the weld has been effective.Another method for testing a butt weld is to hammer the welded panels on the joint into a deeply shaped wooden block to form a deep double-curvature shape.Any sign of a fracture indicates that the weld is not secure.

It is important that all welds are robust, as any weakness or defect in the joints could cause failure in the structure when the vehicle is placed under stress. This is likely to occur when the vehicle is road-tested after being fully restored. **This is the most costly time to have to resolve such a critical issue!**

To become competent at butt welding, a method that requires more practice to master than other types of joint, it is worth having an area set up in the workshop where you can spend time welding with the MIG and TIG. Ideally, this should be a steel bench to which you can easily attach clamps for holding practice pieces of sheet metal.This is also a useful area to carry out a pre-weld test of the machinery and equipment prior to welding on a panel or vehicle body.

Cut out plates of sheet metal of the same thickness that you will be using on the vehicle.A good size for the practice plates is 250 × 100mm (10 × 4in).You could have a large quantity of these cut to size by a fabricator or sheet-metal supplier so that there is a good supply on hand ready for practice or testing of welds. It is best to spend around twenty minutes in practice at one time; any longer will not prove beneficial as you will tire and become less focused, possibly leading to poor results.

CHAPTER FOUR

METALS

To achieve a satisfactory welded joint using any of the welding processes will require some knowledge of the materials used. This is particularly important where the joint will be placed under a degree of stress and is of structural importance to a vehicle bodyshell.

METALS USED ON CLASSIC CARS

A range of metals is used on classic vehicles – most commonly mild steel and alumimium alloy for body panels with brass and stainless steel used for trim panels. More modern vehicles produced from around 1980 onwards will incorporate higher carbon steels that become more difficult to weld and require specialist equipment to carry this out effectively.

USE OF DIFFERENT METALS

Metals are chosen for their required purpose according to the following characteristics:

- strength: dependent on the type of structure of a bodyshell
- durability: resistance to corrosion and fatigue
- malleability: ability to be compressed without fracturing
- ductility: ability to be drawn or stretched out without fracturing
- annealability: ability to be softened to facilitate cold working
- weldability: ability to be welded without a change to the structure of the metal
- conductibility: ability to conduct electricity or heat
- affordability: a main consideration for low-cost mass-produced vehicle bodies.

METAL ALLOYS

Pure metals are comparatively better conductors of heat and electricity than alloys and generally have better corrosion resistance, making them more durable. Pure metals are generally also extremely malleable. However, metals are rarely used in their pure form, being generally relatively weak, so are processed to form alloys that are a mixture of different elements, as this provides greater strength. Pure metals have equal sized grains in their atomic structure, which makes them more malleable or softer. Metals with unequal sized grains in their atomic structure are mixed to form alloys. The unequal size of the grains makes them less malleable or harder. The effect is like grains of sand that when of equal size create a fluid mass; when they are of unequal size they create a solid mass such as used in aggregate to form concrete.

MILD STEEL

The most widely used metal alloy for making vehicle body panels is mild steel, which is an alloy of iron and carbon. It is used because it has excellent malleability, ductility, weldability, inherent strength and is comparatively low in cost. It is easily formed into deep shapes by pressing and is a very weldable metal using all the welding processes available. The only disadvantage is its poor durability, as it is very prone to corrosion when not adequately protected from the effects of atmospheric gases and moisture.

Vehicles from the mid-1970s onwards are generally made from a steel alloy with a higher carbon content, which is far more difficult to work because of its greater strength and also harder to weld because of its thinner gauge and higher carbon content. It is used to give greater strength to the structure and to save weight by allowing bodyshells to be constructed

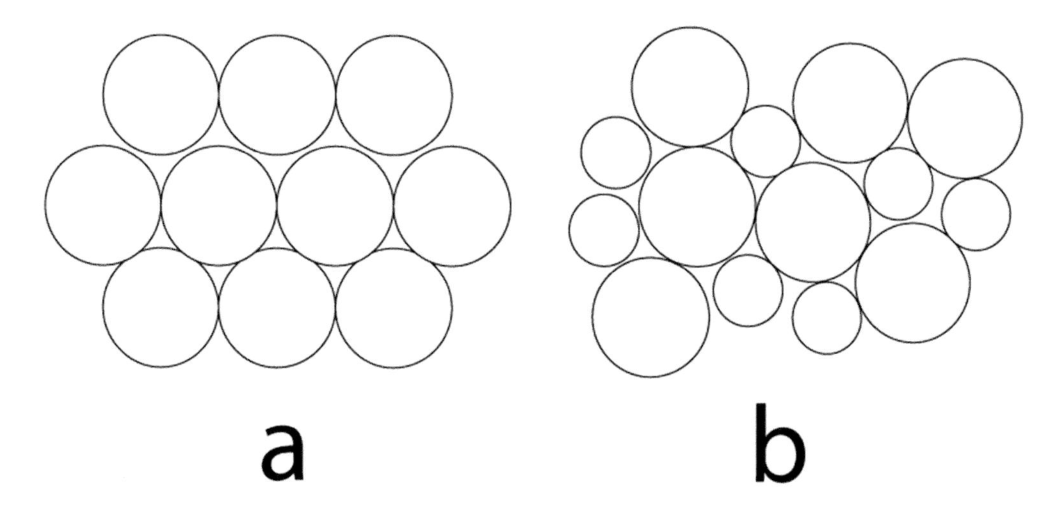

Effects of alloying metals: a) a pure metal has equal sized grains in its atomic structure, making it more malleable and softer; and b) a metal alloy has unequal sized grains in its atomic structure, making it less malleable and harder.

Coach-built bodies are made from aluminium alloy, which is a very malleable material, making it practical to form double-curvature shaped panels by hand.

from thinner gauge material; usually 0.7mm thick. Earlier body panels were made from 0.9–1mm thick material. A higher carbon steel is more prone to corrosion, which is evident on vehicle bodies produced from the 1970s onwards.

ALUMINIUM ALLOY

Aluminium alloy sheet was widely used on early coach-built vehicle bodies because it has a high degree of malleability and ductility when annealed, so making it comparatively easy to form into complex shapes by hand. This made it an ideal material for the coachbuilder to form into complex double-curvature panels by hand using inexpensive tooling. Though it has comparatively little strength, this is overcome on a coach-built body as the panels are supported by a strong wooden frame sitting on a rigid steel chassis.

Aluminium alloy was also used on many post-war vehicle bodies due to its wide availability at the time. In fact, there was a glut of the material following the cessation of production of aircraft immediately after World War II, whilst at the same time steel was in short supply.

The Jowett Jupiter bodyshell, which was developed at a similar time to the Jaguar XK, is made from an aluminium/copper/manganese alloy with the trade name of Duralumin. This was commonly used for aircraft panels due to its comparatively high strength. When annealed, Duralumin is extremely malleable, yet is extremely tough when fully hardened. The main benefit of this particular alloy is that it naturally age-hardens without further working, reaching its maximum hardness over one to two days. This ensures that the finished panels will always be in the fully hardened state when fitted to the vehicle, whereas alloys that do not age-harden may be left partially annealed and so comparatively weak.

Aluminium alloy has seen a resurgence on many modern vehicles, with a push to save fuel and increase performance. Because of its lightness compared to steel (all aluminium alloys are around a third of the weight of mild steel), aluminium alloy offers these benefits, though at a greater material cost. Higher strength alloys are used on modern vehicle bodies that offer similar or even greater strength than mild steel, yet at a third of the weight.

BRASS

Brass, which is an alloy of copper and zinc, has long been used for producing sheet-metal trim sections on vehicle bodies. This is easily formed when

Chromium-plated brass sheet trim panel. Brass was commonly used on older vehicles for trim panels, as it is easily formed into complex shapes and provides a good base for chromium plating.

Materials used in vehicle body construction

The Jaguar XK120 is an example of where different materials were used in body construction. Although the main bodyshell was produced from mild steel, a range of different aluminium alloys were used on certain panels of the body. These are the bonnet, boot lid and doors, which are all opening parts that need to be lighter in weight to facilitate lifting or moving by hand. The lighter weight also places less stress on the hinges than a heavier steel panel would. The alloy used for each panel was chosen for its particular characteristics.

The bonnet was made from a 3000 series (aluminium/manganese) alloy that is of medium strength and can be drawn into a deep double-curvature shaped over a large area. The thickness of material used is 1.5mm (16swg). Although there is little structure supporting the panel, the deep double-curvature shape of the bonnet adds rigidity, making it strong enough structurally to support its own weight.

The door frames and skins were made from a 5 series (aluminium/magnesium) alloy that has a higher strength. It can be formed into double-curvature shapes without difficulty over smaller areas. The door frame is mostly flat sheet metalwork needing minimal pressing. The material thickness used for the frame and the skin is 1.2mm. The complicated structure of the frame combined with the double-curvature shaped skin creates a strong lightweight panel, making it ideal for a vehicle door.

The boot lid skin fitted to the Jaguar XK is made from a very high strength alloy that can be formed into a shallow double-curvature shape. The thickness of the alloy used for this panel is 1.2mm; this is not strong enough in itself, but is supported by a curved laminated wooden frame. The frame combined with the thin skin creates a rigid panel of reasonably low weight.

The Jaguar XK body was built from an extensive mixture of materials. Some parts are timber-framed, while other panels and assemblies are of all-steel construction.

Stainless steel replaced brass on later vehicles as a more cost-effective option because it does not require chromium plating and can be polished to a high gloss finish.

annealed and readily accepts chrome plating. It is comparatively tough when work-hardened and has a high degree of resistance to corrosion.

STAINLESS STEEL

Stainless steel is more prevalent on later vehicles, used for bumpers and trim sections as a more cost-effective option over chromium plated brass as it requires no plating and can be polished to emulate chromium plating. It is extremely tough and highly durable.

SPECIFICATIONS OF METALS

MILD STEEL

Pure Iron is too soft to be useful in most engineering applications so is alloyed with carbon to create steel. Mild steel, which is generally used for sheet metalwork, has a low carbon content of 0.1–0.25%. It is divided into two main groups:

- hot rolled or hot reduced – HR or DD11
- cold rolled or cold reduced – CR or DC01.

Hot rolled mild steel This is rolled whilst hot, producing a material that has a dark blue surface (ferrous oxide layer). This is only produced in thicker plate; the minimum thickness produced in hot rolled sheet is 1.5mm.

Cold rolled mild steel This is first hot rolled, then left to cool. Following this process, it is annealed to soften the material. This makes it more malleable and ductile and easier to weld, making it suitable for forming complex-shaped panels by pressing and joining by welding to form large structures. The metal is descaled following the initial rolling process to leave a bright silvery finish to its surface. It is then rolled again when it is cold to achieve a precise uniform thickness. Cold rolled sheet is commonly sold in various thicknesses from 0.7mm up to a maximum of 3mm.

The most commonly used material for sheet-metal body panels on pre-1980 vehicles is CR4. CR is abbreviated from Cold Rolled and 4 refers to the degree of malleability in the material. Cold rolled steel is produced in a range, from 1 being the most malleable to 8 being the least malleable. CR1 or deep drawing steel is often used for replica (pattern) parts, particularly on deep pressings, as the force required to form the panel in the press is significantly reduced when using the softer material. However, this can cause problems when fitting panels to a vehicle body, particularly on a car without a chassis, as the strength of the complete body will be compromised by the metal that is weaker than that used on the original panel.

The thickness of the material used on different makes and models of vehicle bodies varies. Most pressed-steel vehicle body panels will be 1mm, which is 19swg. Some manufacturers used 1.2mm (18swg) for added strength, but this adds significant weight, or they used 0.9mm (22swg) to save weight, particularly on later vehicles produced from 1970 onwards. Vehicles produced from the mid-1970s onwards used a higher carbon content steel alloy that is more difficult to work and weld.

ZINC-COATED MILD STEEL

Zintec is a zinc-coated cold rolled mild steel sheet often used for remanufactured pressed-steel body panels, zinc having a high degree of resistance to atmospheric corrosion. This allows the panels to be stored for long periods of time without the necessity for any further protection against corrosion. Zintec can be spot-welded successfully, but be aware that the gases produced during the welding process are highly toxic. It is of critical importance to wear a suitable fume mask to protect against breathing in the toxic gases that are given off during welding. The zinc coating should be removed around the joints where MIG or TIG welding is used so as to avoid problems with contamination of the weld pool. This is best done with a fleece flap wheel fitted to a hand angle grinder to avoid thinning the metal of the panel.

LEAD-COATED STEEL

Tin and lead-coated steel were used for various parts on vehicles, mainly fuel tanks, as the coating protected the inside of the tank from corrosion. The coating also made it easier to solder seams as there was no need for fluxing the joint, which would be necessary on a bare steel panel.

HIGH-TENSILE STEEL

High-tensile spring steel and forged components, such as suspension springs and steering linkage arms, should not be welded by the novice, as the welding process alters the temper and strength of the components, making them either too soft or too brittle and liable to fracture. Such parts should always be sent to a specialist welder who can guarantee the integrity of the weld.

ALUMINIUM ALLOYS

A range of different aluminium alloys are used for vehicle body panels depending on the requirements for strength or formability. Aluminium alloys are divided into two main groups – cast alloys and wrought alloys. In making vehicle body panels, only wrought alloys are used. These are further divided into annealable and non-annealable alloys. Annealable alloys can be softened using heat and welded without any change to the structure of the metal, whereas non-annealable alloys cannot be softened and are not weldable without altering the inherent structure of the metal.

Wrought alloys are used for sheet metalwork on vehicle bodies due to their malleability and weldability. Lower strength annealable alloys are commonly found on handmade coach-built bodies, whereas higher strength non-annealable alloys are used on pressed panels on modern vehicles. All aluminium alloys melt at around 660°C, yet have a higher degree of heat conductivity than mild steel. This means that they will require more heat input to melt the metal,

Aluminium alloys used for vehicle body panels

- **1050A** – this grade is almost pure aluminium, not alloyed with any other metal, though it has other elements within it. It has excellent corrosion resistance. It is very malleable and ductile and very weldable. It has low strength, so is unsuitable for structural components. It was traditionally used on panels fitted to a timber frame, as the edges of the sheet can be easily worked around the timber and the panel is supported by the framework to give it strength.
- **3103** – an aluminium manganese alloy that has good malleability and ductility and can be successfully welded. It has reasonable corrosion resistance. It has medium strength and is generally used to form panels that require a degree of working to form deep double-curvature shapes.
- **5251** – an aluminium magnesium alloy that has low malleability and ductility and is more difficult to weld. It has reasonable corrosion resistance. It has high strength, so is more suited to structural components, such as used on a monocoque body. It is also used to form shallow double-curvature shaped panels such as door skins and boot lid skins.

so requiring higher amperage to be used with electric-arc welders.

All of the above aluminium alloys can be supplied in different states of temper (degree of hardness). A code is used to denote the temper of the alloy: '0' is the fully annealed or softest state; 'H14' is half hard (halfway between its softest and hardest state), meaning that it can still take a degree of working; while 'H18' is the fully hardened state that has no malleability, so is only used for flat panel work where maximum strength is required.

Sheet aluminium alloy is usually sold in the H14 (half hard) state of temper, as this allows for further working to be carried out, but is not so soft that it can't be left in this condition in areas of the panel that are not worked.

The thickness of material used for body panels varies on different makes and models of vehicle. Most aluminium alloy panels will be 1.2mm (18swg) or 1.5mm (16swg).

STAINLESS STEELS

Stainless steel is an alloy of steel and chromium. A very thin layer of chromium oxide forms on the surface of the metal and being very hard protects it from atmospheric corrosion. Stainless steel will corrode in the presence of chlorine. It is best used in the polished state to avoid atmospheric corrosion, as any scratches or abrasions in the surface of the metal will make it more prone to corrosion.

Various alloys of stainless steel are available. The two alloys commonly used for making sheet-metal panels for vehicles are:

- 304, which is commonly used to produce body panels and trim sections
- 316, which has a higher resistance to atmospheric corrosion and is generally used for panels that would be used in marine conditions. These are more prone to corrosion due to their proximity to salty water, which acts as a catalyst for the corrosion of metals.

The letter 'L' placed after the alloy number denotes that it is in the annealed condition, giving it a higher degree of malleability. Both alloys have a good degree of malleability and ductility and excellent welding characteristics. They are non-magnetic until work-hardened, which makes them mildly magnetic.

When welding stainless steel, a very hard chromium oxide is produced on the rear side of the weld where the molten metal is not shielded from the atmospheric gases. This is difficult to remove by grinding due to its extreme hardness and so is best avoided. Use a suitable flux or metal backing placed behind the joint to exclude the atmospheric oxygen from the weld and prevent the oxide from forming.

BRASS

Brass is an alloy of copper and zinc. It has good corrosion resistance and is a good base metal for chromium plating that has been proven to last a significant period without any surface corrosion. Available in various grades of alloy, the two common sheet-metal grades are:

- CZ106
- CZ108

Both are relatively low in zinc content, making them suitable for welding. Toxic fumes are given off during the welding process, so a suitable fume mask must be worn to avoid breathing in these gases. Brass is highly malleable and ductile when annealed, yet it becomes very hard when fully work hardened. Repeated annealing is necessary to form any depth of shape into sheet brass, as it quickly work-hardens.

ANNEALING AND HARDENING METALS

All metals used for vehicle body panels will suffer from a degree of annealing during the welding process. This softens the metal, creating a weaker area on and around the weld. This is corrected by hammering the joint after welding to return the metal to its original hardness and strength.

Metals such as mild steel and aluminium are fully annealed (in their softest state) when first produced. They are then partially hardened by rolling to give them a greater degree of strength to make the material more useful for most purposes. Further working by pressing, rolling or hammering will add more strength up to the metal's maximum tensile strength. To work it further (beyond its maximum tensile strength) would cause the metal to fracture. In order to work it further without causing fractures in the metal it requires annealing to soften the material, so restoring its malleability and ductility.

Annealing involves heating the metal to a high enough temperature so that its original crystalline structure is reformed (the recrystallization point), resulting in a softer, more malleable state. This is best done as rapidly as possible and held at the annealing temperature for the shortest time necessary. Annealing naturally occurs during the welding process, so following any welding the joint needs to be work-hardened by hammering or rolling under pressure to restore its hardened state. This annealing will be generated around 25mm (1in) on either side of the welded joint. If the metal is left annealed following welding it will leave a comparatively weaker area around the weld, resulting in the panel deforming in this area when placed under any load.

The effects of annealing are far more pronounced on aluminium alloy than mild steel. Aluminium alloy has around half its maximum tensile strength when in the annealed condition.

It is important to anneal aluminium patch panels prior to fitting in order to avoid the patch altering in size as you are welding it in place. This is due to the expansion and contraction of the metal as it is heated and cools. Aluminium alloy expands 50 per cent more than steel when heated. The effect is more significant on a work-hardened panel, whereas the annealed panel has a greater degree of elasticity so is better able to react to the expansion and contraction that will occur during welding.

Propane is the most appropriate gas to use for the annealing process, as it will reach a high enough temperature to anneal the metal with less risk of burning it than using oxyacetylene. A large burner is required as volume of heat is needed, rather than a concentrated heat source, particularly on a large panel.

ANNEALING ALUMINIUM ALLOY

When annealing aluminium alloy, the cleanest and most effective way to judge the temperature of the metal is to stroke a used matchstick lightly over the surface of the panel repeatedly while you are heating it. It will leave a solid brown mark when the correct temperature has been reached. It is a happy coincidence that a matchstick chars at the same temperature that aluminium anneals, around 420°C. Matches are made from a specific timber that makes them a reasonably consistent test of the temperature reached.

Aluminium is annealed by heating to 420°C. A household match chars at this temperature, providing a good indicator of the temperature reached.

A more general indicator of approaching the annealing temperature is indelible marker pen ink. This is drawn on to the surface of the panel before heating it. As the panel is warmed, the ink turns grey when the metal is getting close to the annealing temperature. Once this is noticed, the metal is then tested with the match, which is a more accurate indicator of the temperature. Take the heat source away from the panel as soon as the annealing temperature is reached. Overheating will result in excessive grain growth due to larger crystals forming at the higher temperature. This will be visible on the surface of the metal as an orange-peel effect. This weakens the structure of the alloy. The metal is cooled following annealing naturally or by quenching in water. The annealing process can be carried out as many times as needed, so as to work more shape gradually into the panel.

ANNEALING BRASS

When annealing brass, the metal is heated till it just turns red. This is best done in a poorly lit workshop, as it is difficult to see the colour change under bright light.

ANNEALING STAINLESS STEEL

Stainless steel anneals at 1,050°C. The metal is heated until red hot to reach the annealing temperature.

CORROSION OF METALS

Galvanic corrosion is the term used for the electrolytic action that occurs between plates of metal in a humid environment. This occurs between metals of the same type, but is more pronounced among different types and grades of metal depending on their atomic characteristics; this is known as bimetallic corrosion. Metals have a place on the galvanic series ranging from the cathodic, which are the least reactive to corrosion, to the anodic, which are most reactive to corrosion.

You will see this reaction on a joint of steel against steel where no protection has been given. The effect is the creation of ferrous oxide (rust) between the juncture of the two plates, which expands as it forms, therefore forcing a gap in the joint that creates further corrosion as moisture is then better able to penetrate the joint.

A similar effect is seen on a joint between aluminium alloys; this will vary depending on the characteristics of the alloy used. Corrosion is exacerbated where different types and grades of metal are in close contact, the most extreme effect being seen on an aluminium/stainless-steel joint, as the difference on the galvanic scale between the two metals is significant. The aluminium will corrode at a faster rate when placed against a more noble (relatively inert) metal. Road salts act as a catalyst to speed up the galvanic corrosion process. Fluxes, which are used in oxyacetylene welding and brazing, will also speed up the process of galvanic corrosion. The effects are more pronounced where there is a mixture of different materials in close contact. Timber frames supporting metal panels will hold moisture that facilitates the process.

Galvanic corrosion is particularly marked on early vehicles that had a positively earthed electrical

The effects of galvanic corrosion are clearly seen on this panel, even though both parts are made from mild steel. The joint is slowly forced apart by the process.

The effects of bimetallic corrosion are exacerbated when there is a mixture of different materials in close contact with timber that holds moisture.

system, as this encourages the electrolytic reaction between the metals, speeding up the process that leads to corrosion. The best protection against galvanic corrosion is to insulate between any lap joint. A weld-through primer can be used on a resistance spot-welded joint. This is a zinc-rich paint that offers good long-term protection. Paint the mating surfaces of the panels prior to spot welding the panels together.

It is also advisable to seal the edges of lapped joints with a body sealant sold specifically for this purpose. Apply the sealant along the edges of all the joints after the panels are welded together. This seals the edge of the joint to protect against ingress of moisture, therefore protecting the joint from galvanic corrosion.

REMOVING OXIDE

All oxides of metal are comparatively very hard, which means that they will blunt the sharp edges of cutting tools. When removing panels from the vehicle, it is best to reserve a set of tools or blades solely for this purpose.

All oxides are best removed from the surface of a panel using a phosphoric acid rather than using abrasives that will thin the metal of the panel, so

Use a zinc-rich weld-through primer to paint the mating surfaces of panels before spot welding together to protect from the effects of galvanic corrosion.

BELOW: ***After panels are welded together use a seam sealer along the edges of all the joints to protect them from moisture ingress and future corrosion.***

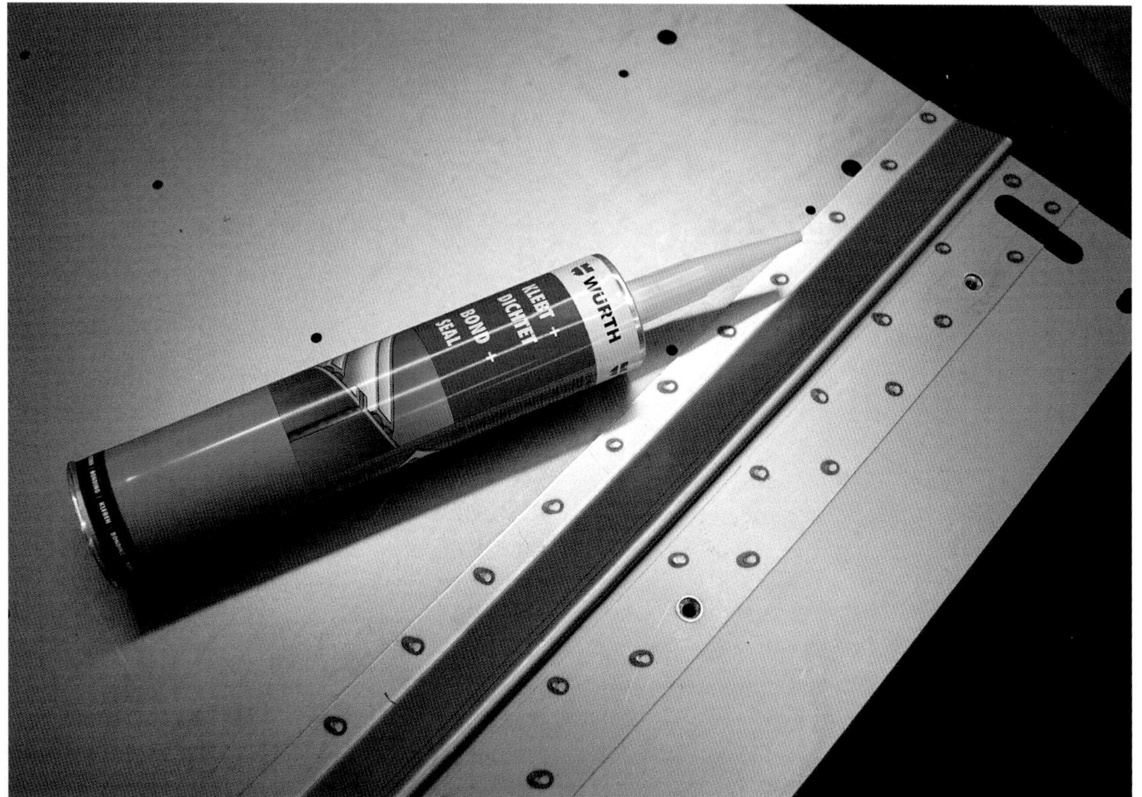

Phosphoric acid is used to clean any surface rust from mild steel. This is simply wiped on and off. Heavier corrosion will require soaking in acid over a longer period of time.

weakening it. A thin oxide layer can be removed by simply wiping the acid on and off the surface of a panel. Heavier layers can be removed by regular recoating or soaking in the acid for a longer period. Always wear acid-resistant gloves when using an acid to avoid acid burns and ingesting the chemical through the skin. Also wear a face shield to avoid harm from splashes on to skin and eyes.

BLAST CLEANING

Shot blasting is effective at removing any oxides without also taking away sound metal, although care must be taken on thin sheet-metal panels, as too much force used in the process will distort the panels. Be cautious when taking panel work to a shot-blasting company, as some will only have experience of cleaning heavy metal structures and can easily damage vehicle body panels. Only use a company that has

Sheet-metal gauges

Standard Wire Gauge = swg

Imperial	Metric	General tolerance
22swg	0.76mm	0.68–0.82mm
20swg	0.9mm	0.83–0.99mm
19swg	1mm	0.93–1.03mm
18swg	1.2mm	1.11–1.32mm
16swg	1.5mm	1.39–1.65mm
14swg	2mm	1.77–2.02mm

As can be seen, there is a wide variation of the tolerance limits for each thickness of sheet metal. It is worth shopping around when buying sheet. Make sure that you determine the actual thickness being sold. **Don't rely on the advertised thickness of the material.**

experience of working on vehicle bodies, as they will understand the correct media and pressure to use to avoid distortion of the panelwork. Immediately following blasting, panels should be protected from further oxide forming by the application of a weld-through primer.

BUYING SHEET METAL

Stockists of metals are divided into ferrous (steel) and non-ferrous (all other metals) to avoid cross-contamination of materials during cutting and storage. Steel stockholders will supply a range of steel-sheet materials, as well as sections of bar, angle and so on. Aluminium stockholders will also supply a range of stainless steels, brass, copper and so on, in sheet form and various sections.

When buying any material, it is important to check its quality. The first consideration is thickness, as sheet metal is often sold at the lower limit of its permissible tolerance. Thinner gauges of metal are also sometimes passed off as a thicker gauge. This can make a significant difference to the strength of a thin sheet-metal panel. A common example is 0.83 material being sold as 1mm. 0.83 is at the thinnest tolerance for 0.9mm (20swg) and significantly thinner than 1mm.

Steel and aluminium alloy sheet should be free from surface oxide. Sheet should be flat, not bowed – this is often a result of the metal being badly stored. Using poor-quality metal sheet will create problems in achieving a clean weld due to impurities in the material and variance in its thickness.

A good source of sheet metal, if you are only using a small amount, is a sheet-metal fabricator, as they will generally be aware of the problems involving quality issues and should be using a good-quality material. They may also have off-cuts of a reasonable size that are adequate for panel repair work.

CHAPTER FIVE

SPOT WELDING

APPLICATION OF SPOT WELDING

Electrical resistance spot welding is the most commonly used method for joining panels on mass-produced steel vehicle bodies due to its efficiency, the comparatively low level of skill required in its operation and the fact that minimal heat is created by this process. The lack of heat dispersion into the surrounding panels avoids the distortion that is commonly created by other welding methods. The majority of sheet-metal panels on most mass-produced vehicle bodies have been designed to accommodate joining by spot welding, using an overlapped joint or square flanges mating at the edges of the panels.

The process is very effective at joining mild steel and stainless steel. Aluminium alloys can be spot-welded, though it requires a high-powered machine to do this effectively. Copper and its alloys cannot be spot-welded due to their high level of electrical conductivity.

THE PROCESS

The electrical resistance spot welding process works by passing a high-amperage, low-voltage current through a copper electrode, copper being an excellent conductor of electricity. The weld is created when a material of higher resistance (usually mild steel) is placed within the circuit. This creates friction, as the current is slowed as it passes through the higher resistance material that then generates heat within the metal at the junction of the two panels. The temperature created by this effect is high enough to melt steel, which is then forged together under pressure, thus creating an effective weld between two or more sheet-metal panels.

List of equipment

- spot-welding machine
- clamps
- protective eyewear – goggles or face shield must be worn to protect your eyes from weld spatter occasionally thrown out during the welding process
- gloves – it is advisable to wear suitable gloves and long sleeves to protect hands and arms from burns
- fire extinguisher – be aware of combustible materials around the welding area, as sparks thrown out from spot welding can create a fire. Keep a hand fire extinguisher close by to put out any flames that may be started.

SETTING UP EQUIPMENT

ELECTRODES

The electrodes (or welding tips) must be kept sharp, clean and accurately aligned to create an effective weld. The diameter of the tip is dependent on the power output of the machine and the thickness of the material being welded. A smaller tip is required on a lower-powered machine to create an effective weld; thicker material will also require a smaller tip to create an effective weld. A rough guide is to keep the tip diameter to a maximum of 3mm for lower-powered handheld machines. For a high-powered machine, the weld tip can be up to 8mm in diameter, which will create a larger area of weld and a stronger joint.

The points of the welding tips can be reshaped initially by hammering, which also hardens the copper. The tips are often annealed due to overheating

The majority of vehicle body panels were originally joined by the spot-weld process, particularly on flat fabricated panels as used on the internal structure of a bodyshell.

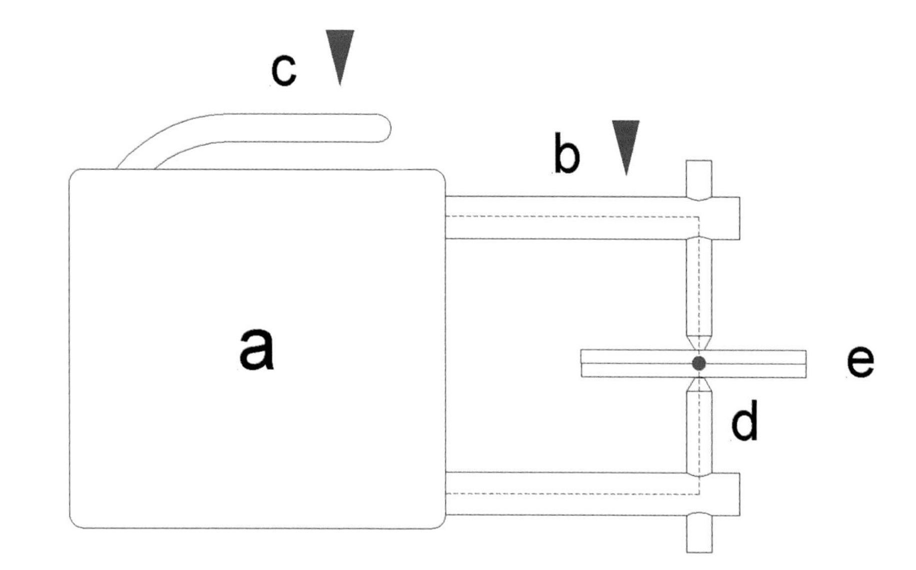

The spot-welding process: a) transformer; b) moveable arm; c) pressure lever; d) adjustable welding tip; e) workpiece.

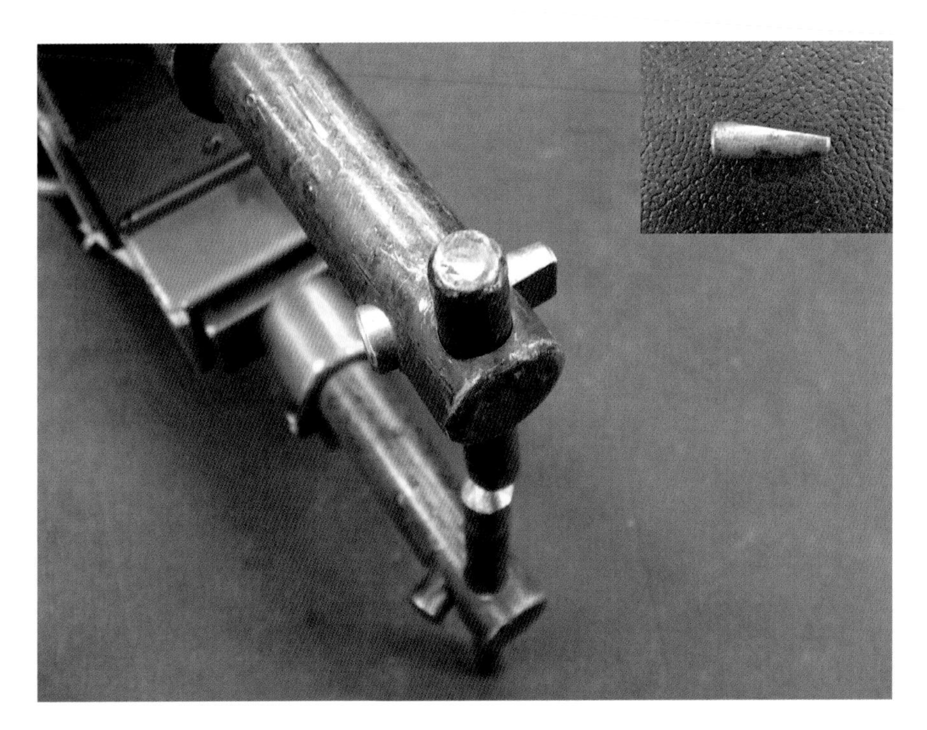

The welding tips are held in place by a tapered pin. The pin is hammered into a hole in the welder arm to lock the tip in the desired place.

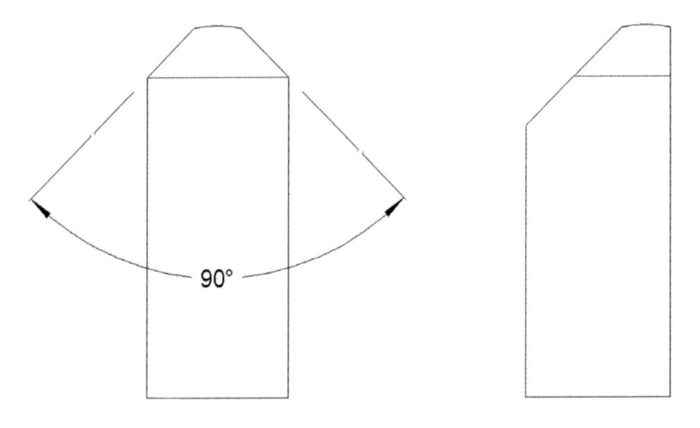

The welding tips are filed to a conical shape with a slight dome to the end face. They can also be filed to an offset to reach into the corner of a flanged edge of a panel.

during repeated use, finish by filing to create a conical point of the desired diameter.

The spot welder tips are commonly fixed into the welding machine's arms with a tapered steel pin. This is pressed into a hole in the arm to lock the tip in place. The tip is released by loosening the pin; use a small hammer to tap the pin out to reposition the tip.

The tips are usually sharpened to a cylindrical tapered point with a shallow dome on the end. They can be sharpened to an offset to allow access into the corner of a flanged joint, where the flange being welded is narrow.

The weld tips must be aligned vertically, directly opposing each other. The welder arms are adjustable, allowing them to be moved in and out from the machine's body and rotated to position them accordingly.

WELDER ARMS

A range of differently shaped welding arms and tips are used to gain access for reaching various configurations of joints on panels. These can be purchased or adapted from those supplied as standard to fit the job being welded. Be aware that a longer arm will

The welding tips must be accurately aligned to create an effective spot weld.

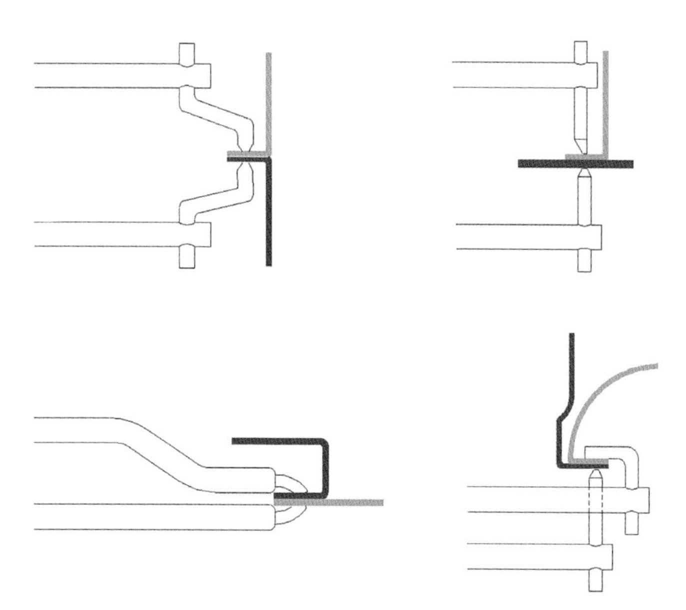

A range of non-standard welding arms and tips are used for reaching various different configurations of joints on panels.

Weld effectiveness

Power output	Material thickness	Tip diameter	Maximum arm length
2KVA	1mm	3mm	150mm
3KVA	1mm	4mm	200mm
	1.2mm	3mm	150mm
6KVA	1mm	6mm	500mm
	1.2mm	4mm	300mm
	1.5mm	3mm	200mm

The figures shown are for welding mild steel. This is a general guide; always test the completed spot weld, as described in Chapter Three.

reduce the effectiveness of the weld. Only use short arms on a 2KVA power machine, as the longer arms will not produce an effective weld on 1mm (19swg) mild steel panels.

WELD EFFECTIVENESS

The table shows the maximum diameter of the electrode tip, the maximum thickness of material that can be joined and the maximum arm length that can be fitted to the machine to create an effective weld, given the power output of the machine.

WELDER OPERATION

The points of the electrodes are placed on a panel and the arms clamped together by use of the handles on the machine to apply pressure to the joint. Beyond the initial clamping, as further pressure is applied to the arms, a micro-switch is activated on the machine that allows the current to flow, creating a weld. This switch is held on for a measured time; some machines have an adjustable timer switch fitted and power will cut out after the set time has elapsed. Machines without a time regulator control rely on the operator to count time, from 0.5–2 seconds maximum. Carry out practice welds on a scrap piece of sheet metal to ascertain the correct time needed to create an effective weld on the material you are joining. Too short a time period will result in a weak weld as the metal between the joint is not melted sufficiently; too long a time period creates unnecessary distortion in the panels being joined, as heat spreads out into the surrounding metal.

Pressure is relieved from the micro-switch to end the current flow after the desired time has elapsed to create the weld, whilst general clamping pressure is maintained on the joint. A degree of pressure must be maintained on the points immediately following welding, so that the molten metal is forged together on the joint as it cools. Keep the welding arms under pressure for a minimum of 2 seconds after the welding current is switched off. **Failure to carry out this procedure properly will result in an ineffectual weld.**

The spot-welder arm pressure is adjusted by means of a hand-turned screw. The degree of pressure that is applied to the welding electrodes is important; if not enough pressure is applied to the electrodes during the welding process the weld will be weakened, as the separate panels will not be effectively forged together. If too much pressure is applied to the welding electrodes, the material will be unnecessarily thinned on the weld, creating a weaker joint. The amount of pressure applied should be just

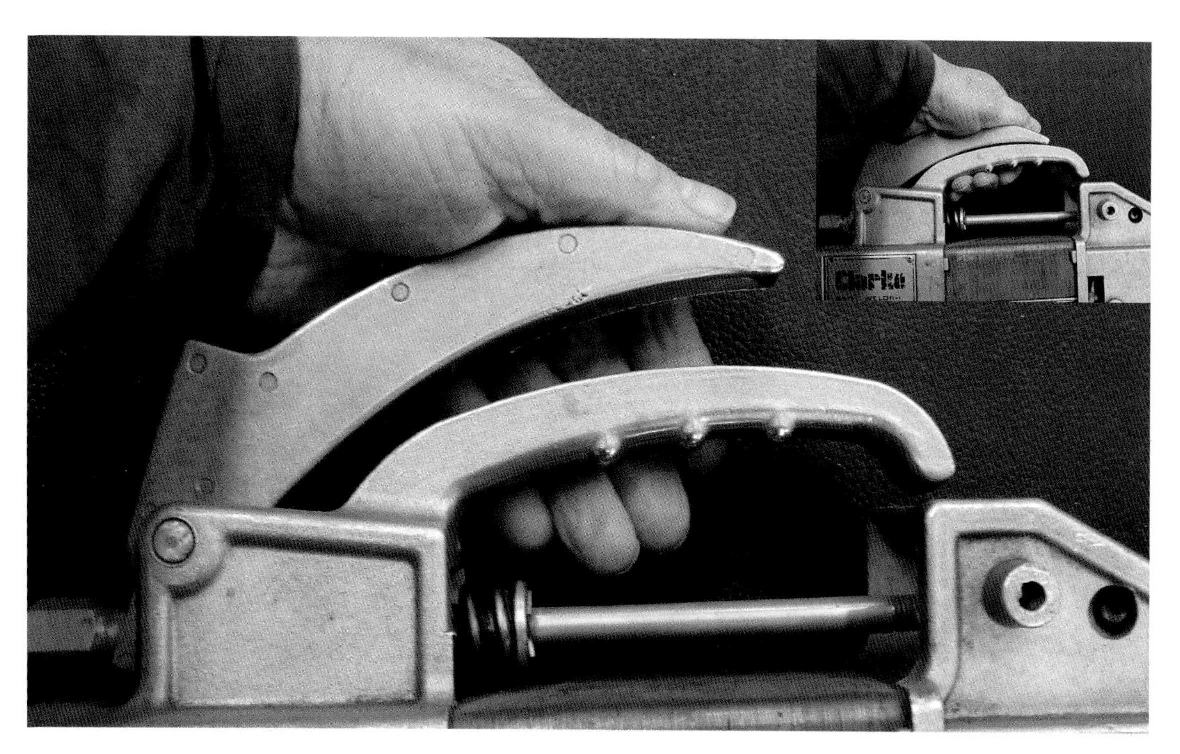

ABOVE: ***A degree of pressure must be kept on the clamping handle immediately following spot welding, so that the joint is forged together as the metal cools.***

The centre of the spot weld should be a bright silver colour, with a small depression left in the panel's surface.

enough to leave a shallow depression in the panel's surface; no deeper than 25 per cent of the material's thickness.

Practise spot welding on a scrap piece of material to acquire a feel for gauging the amount of pressure needed to achieve the desired depression in the panel's surface. Too high a current, too great a pressure and too long a time period for the weld will result in expulsion of molten metal from the weld pool. This is to be avoided, as it will make the metal thinner on the weld and so weaken the joint.

The centre of the spot weld should be a bright silver colour, with a small depression left in the panel's surface.

The strongest weld is achieved by using the maximum power available from the machine, with the shortest duration required to achieve an effective weld. Keeping the weld time as short as possible also minimizes any distortion, as the shorter weld time results in a smaller area of heat spread into the surrounding panel.

AVOIDING INDENTING THE PANEL

In some instances, it is desirable to maintain a level surface on one side of the panel, such as on the outer face of a door skin. Door skins were usually originally spot-welded through the frame at the front and rear edges to avoid movement of the structure. When fitting a new skin, a flat contact head should be fitted to the spot welder to avoid the electrode indenting the panel on the outside face of the door skin. Alternatively, a thick piece of sheet copper (around 3mm [$^1/_8$in] thick) can be placed between the electrode and the panel. The electrical current will flow through the copper plate and create an effective weld without indenting the surface of the panel.

Use a piece of thick (around 3mm) copper plate, or a gimbal spot-weld head, to avoid leaving a depression on the outer face of a panel as shown on the left-hand weld.

SPOT WELDING ALUMINIUM ALLOY

Aluminium alloy requires significantly more power to spot weld than mild steel because the metal is a good electrical conductor and so provides little resistance to the flow of the weld current. A reasonable degree of electrical resistance is necessary to generate enough heat to produce an effective spot weld. This lack of electrical resistance in the aluminium can be overcome by placing a strip of higher resistance metal (steel or stainless steel) on one or both sides of the panels being joined. The higher resistance of the steel strip creates the desired effect of slowing down the weld current to create enough heat within the aluminium to produce an effective weld. Use less pressure on the spot-weld arms when welding aluminium than would be used when welding steel, as too much pressure will result in significant expulsion of metal from the weld pool, creating a weak joint.

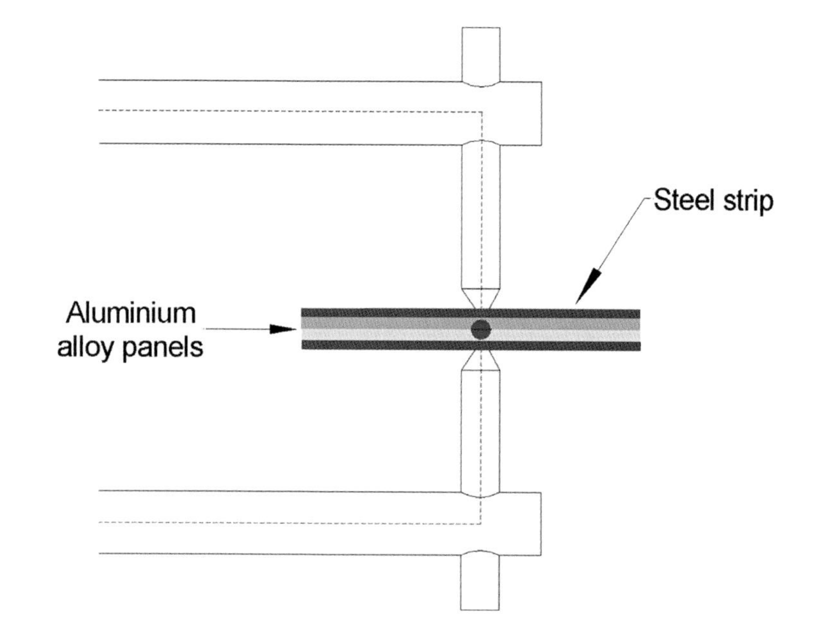

Spot welding aluminium alloy requires a very high-powered spot welder. This can be done using a smaller machine by sandwiching aluminium panels between steel.

Hammer directly on the spot welds to work-harden the metal.

Aluminium alloys will be annealed on the weld following the process, making the metal significantly softer, so it is important to work-harden the metal following any welding. Use a small-headed hammer to ensure accuracy, hammering directly on each spot weld to work-harden the annealed metal.

COOLING

It is important to keep the machine and the electrodes cool during operation to avoid any power loss to the points. Some machines will have a water-cooling facility fitted. Machines lacking any cooling facility should be left to cool occasionally between welds. Heat generated in the copper arms and the transformer results in a significant power loss to the joint, resulting in a poor weld. This is because the electrical conductivity of metals is reduced with the temperature rise in the material. Also, any significant heat generation at the welding tips will result in the copper being annealed (softened), causing the points on the welding tips to become blunt. These can be hardened by hammering the points back into shape before filing to the correct diameter.

JOINING PANELS

PREPARING THE JOINT

The metal needs to be clean and free from rust, oxide, paint or oil contamination. A common problem on older vehicles where paint protection was poor is electrolytic or galvanic corrosion between the joints of panels. This causes the joint to be pushed apart between the spot welds by the oxide layer that is formed between the mating surfaces of the metals, resulting in a quilting effect along the joint. The effect is aggravated, as when the panel surfaces are pushed apart moisture can reach further into the joint, creating additional corrosion.

A phosphoric acid, such as Jenolite, can be used to clean the surface of metal prior to welding. This is simply wiped on and off, leaving an iron-phosphate finish to the surface of the metal. This is mildly corrosion resistant and provides a good key for applying paint primer. A zinc-rich weld-through primer can be applied to the mating surfaces of the joints being welded to avoid any future corrosion occurring in-between the joint.

Galvanic corrosion will create a quilting effecting, pushing the panels apart. To avoid future corrosion, paint the mating surfaces of the panels with a weld-through primer.

CLAMPING THE JOINT

Clamp both panels being joined securely, using locking pliers if possible. Self-drilling, self-tapping screws are also good for this purpose, as they will pull the panels tightly together. Always ensure that the panels are a good, accurate fit. The joint must be tight to achieve an effective weld. **Do not rely on the pressure applied to the spot-welding machine arms to pull the joint tight together.**

PLACEMENT OF WELDS

Always test the effectiveness of the spot-welding machine on a scrap piece of sheet metal prior to welding any panels together, as described in Chapter Three.

Tack both ends of the panels first, then the middle of the joint, gradually working in-between previous spot welds until enough welds are in place to create a strong joint. The spacing of the spot welds should be

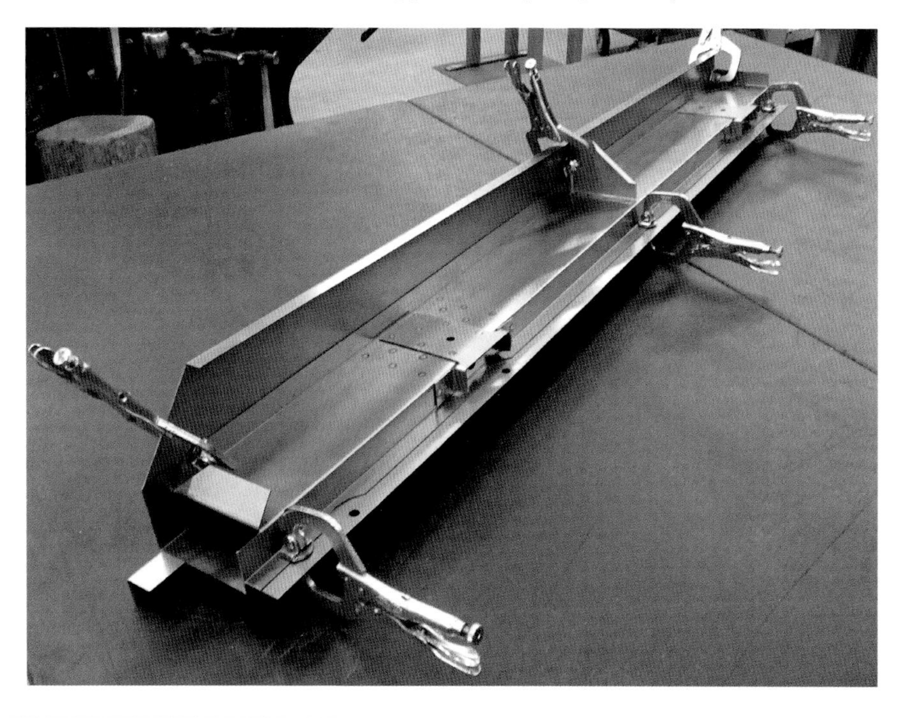

Clamp panels together securely to avoid the clamping force from the arms of the spot welder pulling the joints out of alignment.

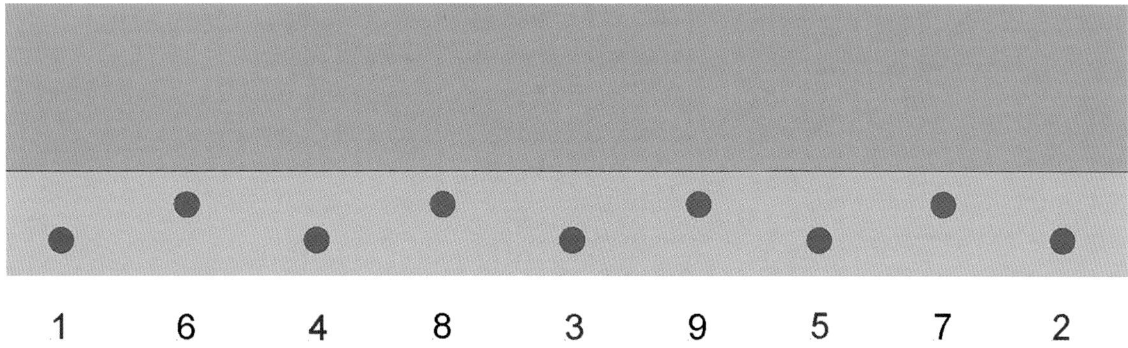

The order of placing the spot welds is important to holding the panels in place initially and avoiding a localized build-up of heat into the panels.

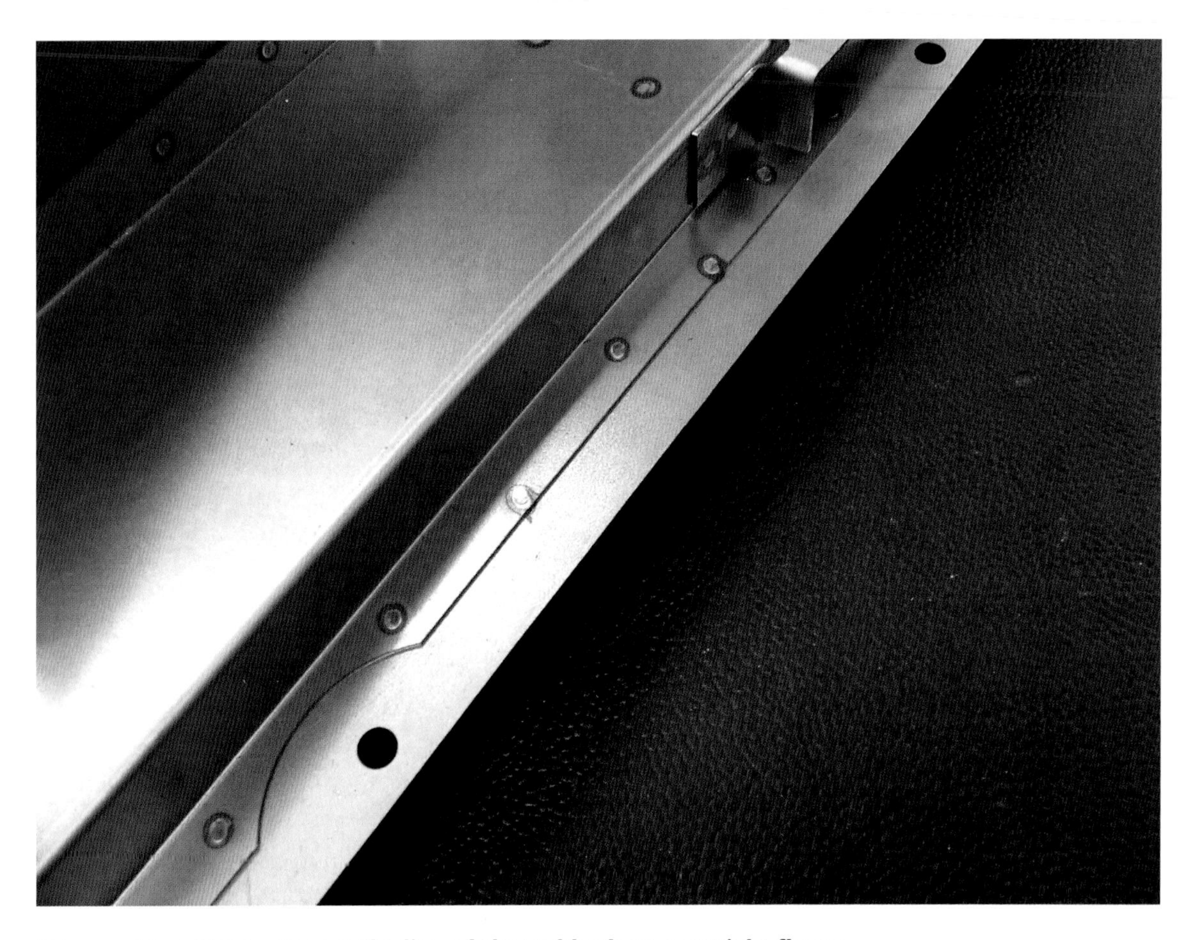

It is good practice to stagger the line of the welds along a straight flange to join a greater surface area. This avoids the panels rocking on the line of spot welds and creates a stronger joint.

a maximum of 25mm apart and a minimum of 10mm apart. If the welds are placed too closely together, the current can pass through an adjacent weld during the welding process, resulting in the weld being ineffective.

The spot welds should be staggered in a zigzag manner, even on a narrow flange, rather than placed in a straight line to provide a stronger joint. Under any stress placed upon it a panel can bend around a straight line of welds, eventually breaking the joint apart. A staggered line of welds avoids this possibility and so creates a significantly stronger joint.

MINIMIZING AND RELIEVING HEAT DISTORTION

Avoid a build-up of heat in the panels being joined by welding in stages and allowing the metal to cool in-between groups of spot welds. Keep the electrodes cool by resting between welds.

Any distortion created by the welding process can be relieved by hammering directly on the spot welds to stretch the metal back to its original size. Do this in as consistent a manner as possible to achieve a smooth finish. Allow metal to cool fully before deciding whether the joint requires further hammering. This action also work-hardens the metal that will have been annealed during welding. It is critical to the strength of the weld, particularly on aluminium alloys, to carry out this process, as the material will be significantly softened due to annealing caused by the heat produced during the welding process.

FITTING A REPAIR PANEL

Spot-welded joints are overlapped, flanged or joggled. A joggled or stepped joint is commonly used when fitting a door repair skin.

Place the initial welds at each end of the joint on the panel, followed by a spot weld in the centre of the length of the joint. Fill in-between these initial welds with further spot welds, allowing time for the machine and the panel to cool in-between welds, so as to avoid a build-up of heat in the metal that would create distortion in the panels.

REMOVING AND REPAIRING SPOT WELDS

REMOVING SPOT WELDS

To remove original panels that have been spot-welded, or to remove a misplaced panel, drill out the spot-welded area on one panel. There are proprietary spot-weld removal drills available for this purpose. These are quickly blunted by drilling into metal oxide, which is extremely hard and often present on an old panel. Proprietary spot-weld removal drills are difficult to sharpen due to their complex shape and so tend to have a short life, making them an expensive proposition.

Alternatively, a standard twist drill bit can be used for this process. Initially use a 3mm ($^{1}/_{8}$in) diameter twist drill bit; drill into the material to the depth of the point of the drill bit that is around the thickness of one panel. Then grind the end of a larger twist drill bit (6–8mm in diameter, depending on the size of the spot weld being removed) to a very shallow angle, creating an almost blunt end. The larger drill bit will follow the initial drilled cone created by the smaller drill bit, but will not go any further than the depth of it without applying significantly more pressure on the drill bit. This will remove the spot-welded area from just one panel, so releasing the joint between the two panels. The hole can then be plug-welded using a MIG welder (see Chapter Six on MIG welding) on reassembly of the panels.

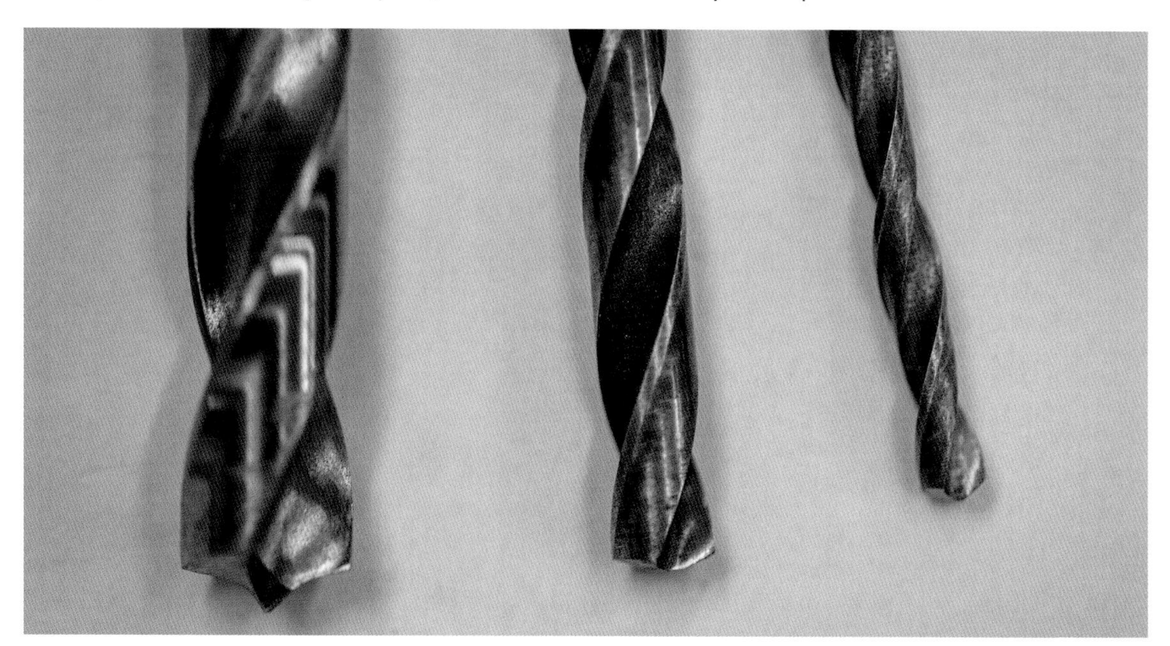

Proprietary spot-weld removal drills are available, but tend to blunt and are difficult to sharpen. Alternatively, use a standard twist drill sharpened to a shallow angle for the same purpose.

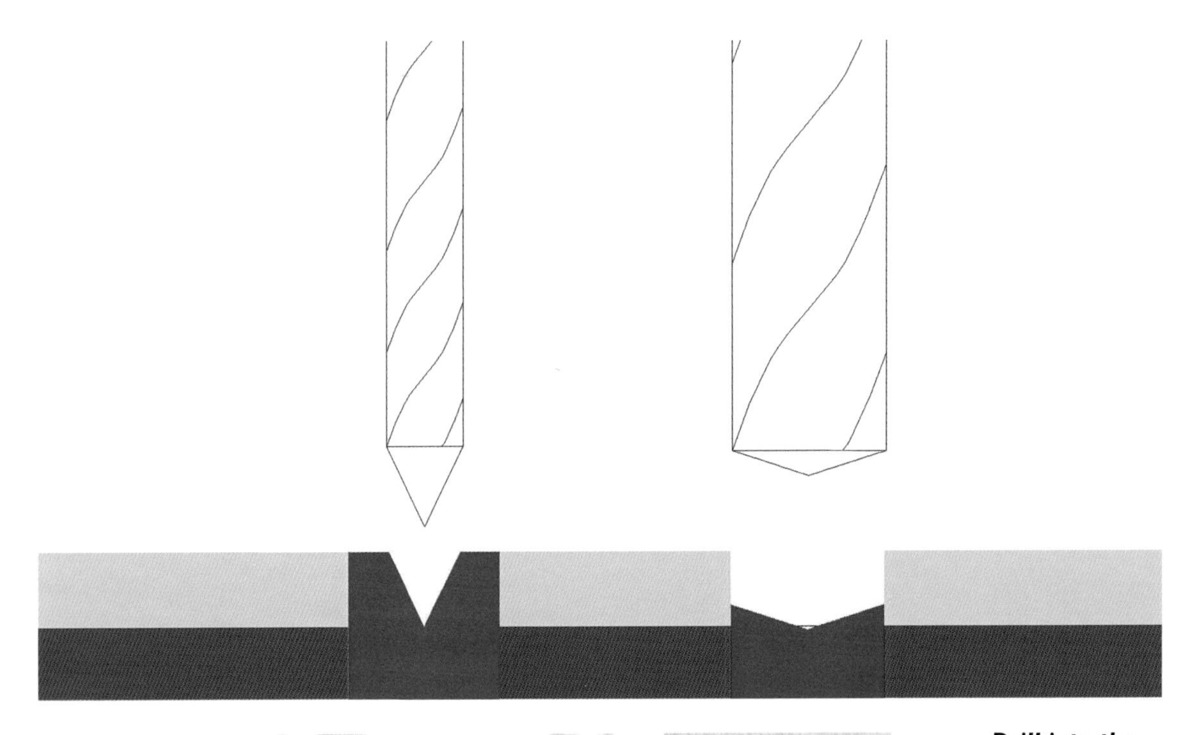

ABOVE: ***Drill into the centre of the spot weld with a small diameter standard twist drill, followed by the larger drill bit, which is sharpened to a shallow angle. This removes the spot-welded area from one panel.***

The spot welds are removed on the outer panel to release the panels for realignment. The holes created are filled with a plug weld using the MIG welder.

The right-hand panel has been cleaned with phosphoric acid to remove the blue surface oxide created by the welding process, prior to painting.

REPAIRING HOLES

To fill a hole with a solid plug of metal using a spot welder, rivet a metal plug of the same material and diameter tightly into the drilled hole. Use a small-headed hammer to ensure accuracy of the hammer blows. Hold a suitably shaped dolly directly behind the hole in the panel to hammer against. Spot weld on top of the plug to secure it in place.

FINISHING THE WELD

Using a small-headed hammer to ensure accuracy, hammer directly on each spot weld to harden the annealed metal. If possible, hammer the joint smooth while it is still hot; any slight distortion that is left will disappear as the metal cools and shrinks.

Allow the metal to cool fully before checking for any distortion left in the panel. Use a phosphoric acid to clean the metal; this will remove the blue oxide produced during the welding process and provide a good key for paint finishing.

The phosphoric acid is simply wiped on and off, leaving an iron-phosphate finish to the surface of the metal. This is mildly corrosion resistant and provides a good key for a primer. A body sealant should also be applied along the edges of any lap-jointed seams to avoid water ingress into the joint. The better-quality sealants are also a structural adhesive, providing a permanent bond with the metal.

CHAPTER SIX

METAL INERT GAS – MIG

APPLICATION OF MIG WELDING

Metal Inert Gas (MIG) welding is commonly used for repairs to vehicle bodywork due to its ease of use and the relative lack of skill required to carry it out. Although MIG welding is commonly used to join butt-welded joints when fitting repair panels to vehicle bodies, it is not the best process for this as it creates an excessive spread of heat into the surrounding panel, resulting in significant distortion. It is best used for fillet or buttress welds, when joining supporting brackets to the structural members of the body or chassis. It is particularly useful in restoration work, for creating plug welds that replicate original electrical resistance spot welds.

Mild steel and stainless steel can be successfully welded using the MIG process with a basic machine. Aluminium alloy and copper and its alloys can be MIG welded, though this requires more specialist equipment.

The MIG welding process is best used for joining heavier brackets to the bodywork or chassis.

THE PROCESS

The MIG process works by striking an electric arc between a continuously fed filler wire and the surface of the panel being welded. The wire acts as an electrode to carry the electrical current to facilitate the welding process. This creates enough heat within the metal to melt the filler wire and the metal at the point of the arc's interface with the panel. At the same time, a shielding gas is fed around the weld pool to avoid the inclusion of atmospheric oxygen and nitrogen into the weld. The filler wire used is a hard steel alloy containing silicon and manganese, which both add certain desirable properties to the metal to produce a clean, strong weld.

List of equipment

- MIG welder and torch
- argon/CO^2 gas
- torch holder – suitable for holding MIG torch securely
- MIG welding helmet – fitted with a light-reactive self-darkening lens
- small-headed hammer – flat polished face
- wire brush – stainless-steel bristled for cleaning aluminium, brass or stainless-steel welds
- welding stool – with adjustable height and tilt mechanism
- dolly
- handheld angle grinder.

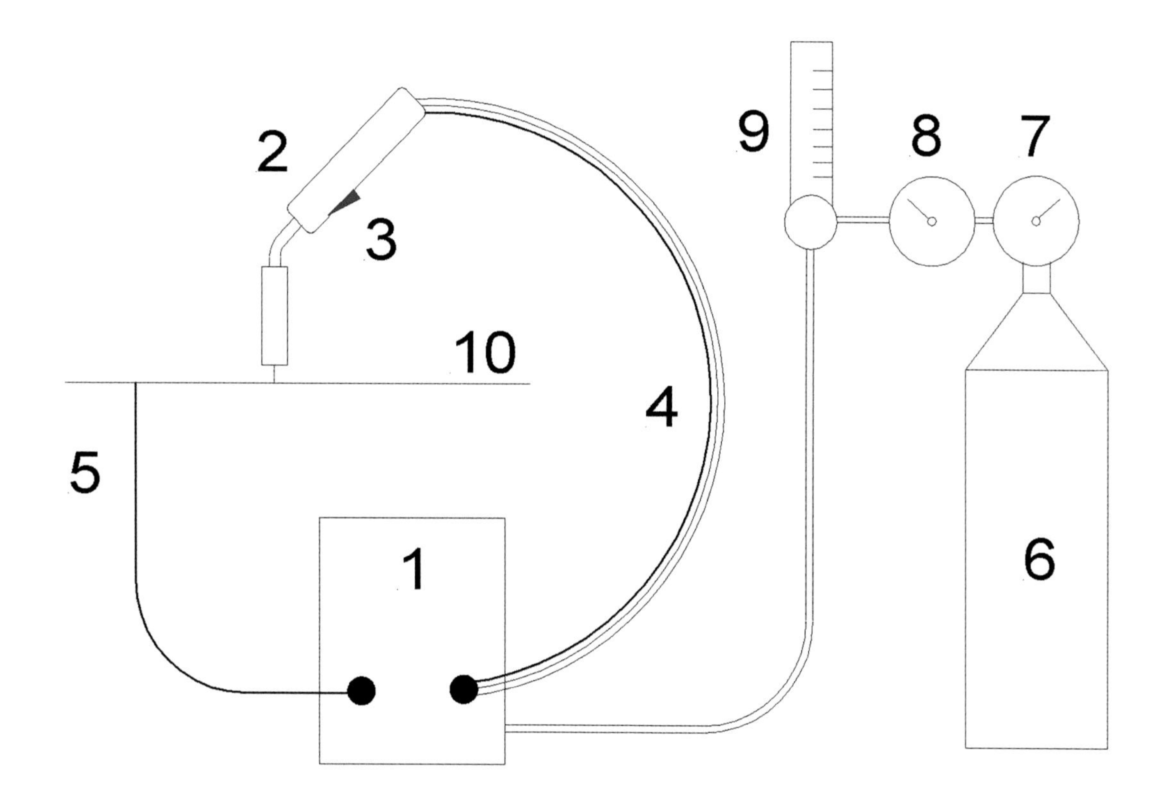

MIG welding process: 1) welding machine – provides DC current supply; 2) torch – feeds wire electrode and directs gas flow around weld pool; 3) switch – controls wire feed and gas supply to torch; 4) torch cable – supplies power to electrode and gas supply through torch; 5) earth lead – connects power supply to welder; 6) gas supply – pressurized bottle containing argon/CO^2 gas; 7) gauge – shows contents of gas bottle; 8) regulator – controls pressure released from bottle; 9) flow meter – controls flow rate of gas supplied to torch; 10) workpiece – metal being welded.

SETTING UP EQUIPMENT FOR 0.9–1.2MM MILD STEEL

WELDING CURRENT

Set at around 60amp for tack welding, 50amp for running-in the weld. Refer to your machine's handbook to show the amperage of the numbered settings.

GAS

The gas flow rate is set at 5–8ltr per minute (lpm).

WIRE FEED SPEED

This is set in relation to the amperage being used. Too fast a wire feed speed will result in the torch being pushed away from the work. Too slow a wire feed speed will result in the wire burning back into the torch nozzle. On some machines this will be automatically set to correspond with the amperage setting.

BEFORE STARTING

Check that the wire feed switch is set to the diameter of the wire that is installed in the machine, usually either 0.6 or 0.8mm. Check that the welding torch contact tip is of the correct diameter to match the wire being used. When first switching on the machine always feed a length of wire through the torch cable by pressing and holding the trigger for two to three seconds; this purges any atmospheric gas that is contained within the torch lead, ensuring that the shielding gas is available at the nozzle as soon as you start the welding process. Cut back the wire protruding from the end of the torch tip so that only 3–4mm is showing before beginning to weld.

TESTING THE WELDER

Run a short weld on a test piece of scrap sheet metal to check that the weld produced is clean and has enough degree of penetration to ensure that both pieces of metal are joined effectively. Check that the

Ensure that the welding wire is protruding from the end of the weld tip by 3–4mm before beginning to weld.

There will be a build-up of weld spatter within the torch gas shroud during welding.

Remove the weld spatter within the torch gas shroud using a steel drift pushed through from behind.

wire feed speed is correct for the amperage setting. The weld should be left slightly raised in profile. Check that the gas is shielding the weld sufficiently; no brown discoloration should be visible around the weld and the surface of the weld should be silver blue in colour.

The welding machine needs to be earthed effectively to the workpiece to achieve an electrical circuit to create a weld. Clean all paint from the surface of the panel on and immediately around the area to which the earth clamp is fitted to ensure a good earth connection.

CHECKING THE TORCH NOZZLE

The torch nozzle needs to be cleaned out regularly, as there will be a build-up of weld spatter on the inside of the nozzle during use that will impede the gas flow, resulting in a poor-quality weld.

It is useful to have a steel drift kept near the machine for this purpose. Regularly remove the nozzle from the torch and push the drift through from the back of the nozzle to remove any weld spatter.

Ensure that the joint is clamped securely to avoid any movement in the panels as they are tack-welded.

CHECKING THE WIRE FEED

Check that the wire feed is working smoothly. The steel welding wire is copper-coated to protect it from corrosion. This copper coating rubs off in the torch liner, resulting in a build-up of copper dust over time, eventually causing the wire to stick and move erratically, creating a poor weld. Occasionally pull the wire back through the torch lead and blow compressed air through the liner to clear the dust. **A smooth wire feed is essential to create a consistent weld.**

PREPARING THE JOINT

The metal needs to be clean and free from rust, oxide, paint, or oil contamination. Joints should be a good, tight fit with no visible gaps. Any gaps in the joint will create a significant amount of distortion due to the addition of molten metal, which shrinks as the weld pool cools. **When making a butt joint it is essential that the panels are cut accurately to provide a tight joint with no gaps.**

WELDER OPERATION

Always test the welder on a scrap piece of sheet metal to ensure that everything is working correctly before attempting to tack or weld on body panels. A clear view of the weld pool is critical; use reading glasses if necessary, to enable you to see clearly at a short distance. Your eyes need to be around 30–40cm from the weld pool. If you wear glasses, avoid bifocal or varifocal lenses while welding as you will need to tilt your head as you progress, which will result in the weld going out of focus. Use single focus reading glasses, if needed, to avoid this problem. A work light aimed at the joint can help to provide a clear view of the weld pool. Lean your

Position the end of the torch nozzle 3–4 mm from the surface of the panel being welded.

upper body forward and turn your head sideways, looking along the joint so that you can clearly see the joint and the weld pool.

Position the end of the nozzle of the torch around 3–4 mm from the surface of the panel. The torch shroud can be initially rested on the panel with the torch inclined away from the operator at a slight angle to the vertical. Once the weld is started, lift the torch so that the shroud is 2–3mm from the surface of the panel and the torch is close to vertical to the panel's surface.

When reaching the end of the weld, the natural tendency is to lift the torch away from the panel when releasing the trigger. **It is critical to avoid this,** as it results in an inclusion of atmospheric gases into the weld pool because the shielding gas has been removed while the metal is still molten. This is a very common fault and practice is needed to train yourself to lift your finger off the torch trigger without lifting the torch away from the weld pool.

Most welding machines will automatically have a post-flow rate of 1 second. This is the period for which the gas will continue to flow after the weld current is stopped. This helps to shield the molten weld pool from the atmospheric gases as the weld cools. The post-flow period can be set for a longer time when welding thicker materials; 1 second is adequate when welding thin sheet metal.

TACKING A LAP JOINT

Clamp the panel or bracket being welded securely in place. Begin by tacking (making a small circular weld around 4mm in diameter) at each end of the panel. Next, place a tack in the middle of the length of the joint and proceed by tacking in-between the preced-

On a lap joint initially tack both ends of the joint before placing further tacks in-between to hold parts securely in place.

Tacks should be around 3mm in diameter and placed at intervals of around 40mm.

ing tacks. Tacks are best placed around 40mm apart. This secures the joint in preparation for running-in (continuously welding) the joint. Sheet metal will distort as you build heat into a panel, so to avoid heat build-up tacks need to be spaced along the joint, allowing time for the panel to cool in-between further tacks. Release the trigger as soon as the weld pool forms to keep the tacks small.

Tacks should be kept to a maximum of 4mm diameter on 1mm thick steel. A larger tack than this will create an inclusion (gas bubble) on the rear of the weld due to a lack of shielding gas on the back of the joint. Ensure that the panel is tacked adequately to hold the joint tight and level along the complete length of the joint.

Grind the tacks level with the surface of the panel, before running-in the weld, using an 80grit flap disc fitted to a hand angle grinder. This will also remove the surface oxide; this is necessary to prevent any inclusion of oxide forming in the weld pool during the running-in process.

TACKING A BUTT JOINT

Clamp both panels together at one end only and work from the clamped end, placing tacks around 25mm apart. This differs from welding heavier plate or sections that would be tacked at both ends first. Sheet metal will distort as you build heat into a panel, so it needs to be tacked progressively along the joint to maintain a level and tight joint. The technique is similar to closing a zip, in that as you work along the panel it will pull in, keeping the joint tight and level as you progress. Use your non-torch hand, holding the panel, to control the level of the joint at the position you place each tack. Grind the tacks level with the surface of the panel before lightly hammering the tacks, against a dolly that is held behind the panel, to level the joint before running-in the weld.

Grind the tacks level with the surrounding panel prior to running-in the joint.

RUNNING-IN THE JOINT

TORCH CONTROL

Controlling both the speed of traverse and distance of the tip of the torch nozzle from the panel is critical to ensuring a consistent smooth finish to the weld. To maintain a steady movement, rest your torch hand in your other hand, moving your upper body with the torch to maintain a constant focal distance from the weld pool. For a right-handed person, position yourself to the left of the torch looking along the joint. Lean the torch away from you at a slight angle to give a clear view of the weld pool. Leaning the torch at too great an angle from 90 degrees to the face of the panel will mean the gas is not shielding the weld pool adequately, so make sure this is no more than 10 degrees from the vertical. Begin the weld at the

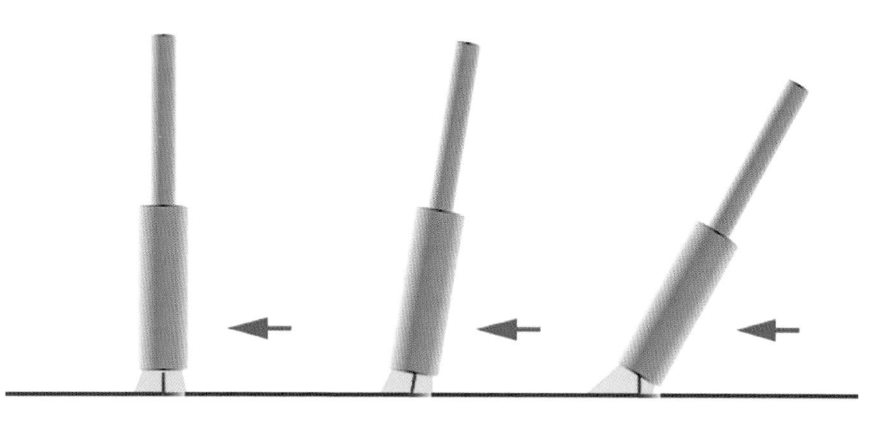

It is important to avoid holding the torch at too great an angle from the vertical, as the shielding gas will not be covering the weld sufficiently.

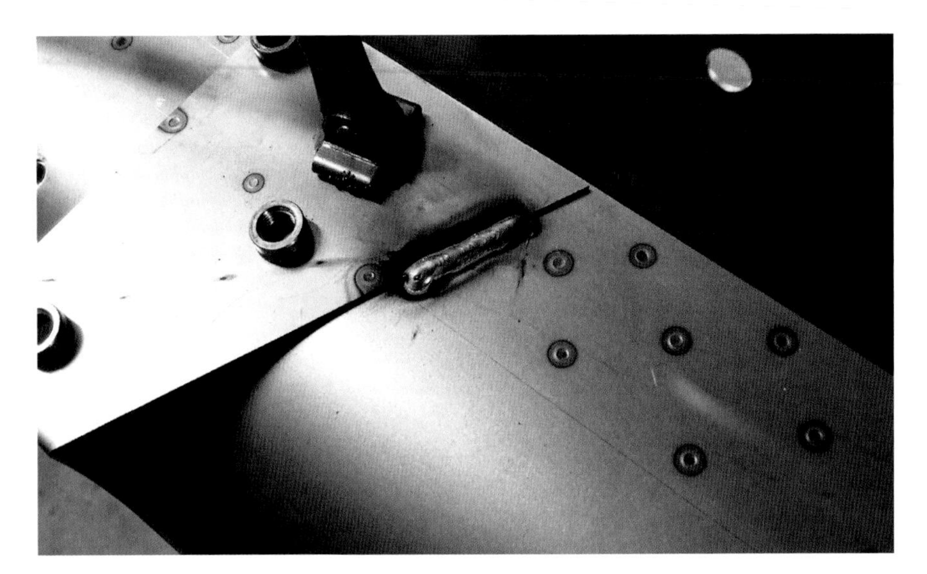

The weld should be around 3mm wide and as consistent a width as possible over its full length.

On a MIG butt weld, the added metal from the process is used to melt the metal of the two panels being welded, fusing them together, without introducing the harder MIG wire into the joint.

furthest point from you, moving towards yourself as you progress along the joint.

For a left-handed person, position yourself to the right of the torch, moving the torch towards you. Always move the torch towards you so that you have a clear view of the joint you are fusing together. Use a welding stool if possible, so as to steady your body as you weld, as any instability in your stance will create an erratic weld.

WELD WIDTH

The aim is to maintain a consistent width of weld. The speed of the traverse will vary depending on the tightness of the joint and will need to be faster as more heat builds into the panel. The width of the weld needs to be around 3mm to achieve enough penetration of the panel, whilst avoiding the weld dipping on its surface. The surface of the weld should be left slightly raised, with visible penetration through the back of the joint. It is desirable to run as long a weld as possible in order to minimize distortion around the start and finish of the weld. Maintain a comfortable position to keep the weld smooth; as soon as your position becomes uncomfortable, stop and move your body to start a new run.

When butt welding using the MIG process, it is desirable to introduce as little as possible of the steel filler wire into the joint. This is because the filler wire is generally a harder material than the metal of the panels being joined and this inconsistency in hardness will create problems when trying to planish the joint smooth. To minimize the addition of any filler metal into the joint the aim is to use the MIG solely as a heat source to melt the edge of the panels being joined, so fusing the two panels together. Any addition of metal should be left proud (on top of the joint) and is ground off to leave the fused joint without any hard filler wire in it. The joint should ideally be placed on the curved surface of a panel to avoid the weld sinking, making it difficult to grind to a smooth finish.

FINISHING THE WELD

The completed weld is initially ground level with the surrounding panel to remove excess metal added by the addition of the filler wire. The joint is then hammered smooth to relieve any shrinkage created by the welding process; this also removes any distortion in the panel. Use a small-headed hammer to ensure accuracy in hammering directly on the weld to flatten and stretch it, with a suitably shaped dolly held firmly behind the joint. Over-hammering on the weld will create new distortion, as the metal then becomes too stretched. So be conscious of the need to stop hammering as soon as the panel becomes smooth.

The weld will be left slightly raised from the surface of the surrounding panel. This is then ground or sanded back level with the surrounding panel's surface. This enables the removal of small defects in the welded joint as further metal is removed. If possible, hammer the joint smooth while it is still hot; any slight distortion left will disappear as metal cools and shrinks. Allow metal to cool fully before checking for distortion. Lightly sand the joint to remove high spots before further hammering and finally sanding to achieve the desired finish for paint preparation.

PLUG WELDING

The MIG welder can be used to emulate a resistance spot weld where it is impractical to use a resistance spot welding machine, such as in the centre of a floor section on a vehicle body where very long arms would be needed to reach the joint. This is usually due to the limitations of the equipment available to the restorer; large spot weld machines are too costly to purchase, are very bulky for use in a small workshop, and they usually require a three-phase electrical supply that may not be available.

PREPARATION

To create a plug weld, first drill a series of 6mm ($\frac{1}{4}$in) diameter holes, spaced at 20mm ($\frac{3}{4}$in) to 30mm (1$\frac{1}{4}$in) apart, on the centre line of the joint in the uppermost panel. Deburr the back face of the

A purpose designed locking plier clamp is available for holding the panels in place around the punched hole whilst plug welding the panels together.

holes before clamping both panels tightly together. A purpose designed locking plier clamp is available for holding the panels in place around the punched hole. This has a copper backing plate to avoid the weld adhering to the clamp.

PROCEDURE

After the preparation, the holes are filled with a short burst of weld. This burst needs to be of a long enough duration to fill the hole with enough amperage to melt both panels together around the perimeter of the hole. A torch shroud or nozzle with two prongs protruding from the end is available for plug welding. This holds the torch the correct distance from the surface of the panel during the process and keeps it static.

The prongs on the end of the spot-weld nozzle hold the torch the appropriate distance from the surface of the panel while the operator pulls the trigger on the torch to create a plug weld. This should be carried out using the maximum power setting of the machine to gain sufficient penetration through the lower panel. It should also be with the shortest duration required in order to avoid heat spread into the surrounding panel. The aim is to leave a slightly hollow dished surface to the weld that replicates the appearance of a factory resistance spot weld without the need for any grinding

Ensure that the torch is held vertical during the process. Some machines will have a timer and a specific power setting for this purpose; if these are not available use the maximum power setting of the machine and hold the trigger on for the required time it takes to achieve a slightly dished weld with visible penetration to the underside of the lower panel being welded. Distortion of the panels is minimized by using a short weld duration. Practise on a scrap piece of sheet metal to gain an understanding of how long it is necessary to hold the trigger before releasing it to stop the weld.

This method can also be used where original resistance spot welds have been drilled out to remove the panel to accommodate repair work. Following repairs, the panel is replaced over the original joint

A gas shroud or nozzle with the addition of prongs is used for MIG plug welding. This holds the torch the correct distance from the surface of the panel during the process.

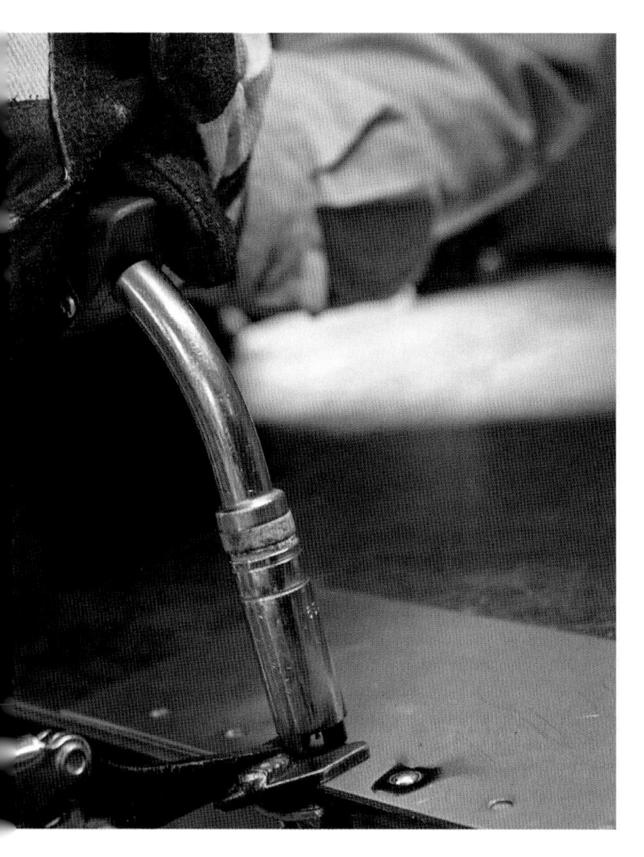

Aim the MIG welding wire into the centre of the punched hole. Position the locking plier to assist with this. It is important to keep the torch held vertically to the surface of the panel.

and the holes plug-welded to replicate the position and appearance of the original resistance spot welds.

MIG WELDING ALUMINIUM ALLOY

The MIG welding of aluminium alloy requires a more specialist machine designed specifically for this purpose. MIG welding is only used for the welding of thicker metals (2mm plus), such as used on brackets and framework. A higher amperage is used due to aluminium's excellent conductivity, meaning that heat is quickly dissipated over the panel. Also, the wire feed requires a finer control, as the soft aluminium wire used in the welding process tends to buckle in the torch lead. Some machines have a motor fitted near the torch to pull the wire through the lead rather than push the wire, which is the more common method of feeding wire, particularly when welding steel panels. The process of welding joints is the same as for welding steel, as explained above, though more time in practice will be needed to set up the amperage and wire feed controls to work effectively on the thickness of metal being welded.

MIG WELDING STAINLESS STEEL

The MIG welding of stainless steel is similar in principle to welding mild steel. A stainless-steel filler wire is fitted into the machine. Argon/CO^2 gas is used for shielding the weld pool.

Stainless steel flows readily when molten. It requires less heat input than mild steel to melt the metal due to its comparatively poor heat conductivity.

A problem experienced when welding stainless steel arises from the scale formed on the rear of the weld. This is chromium oxide, which is extremely tough and difficult to remove, even with a mechanically powered grinder. To avoid the build-up of scale, make sure that the weld does not penetrate through the joint more than is necessary. A flux can be used, applied to the back of the joint prior to welding. This prevents atmospheric oxygen being introduced into the weld pool. Another tactic to prevent the oxide forming is to back the weld with a strip of metal to prevent the atmospheric gases reaching the weld pool.

STITCH WELDING

A good general use of the MIG welding process is the joining of two panels that meet squarely as a butt joint, by welding on the external corner of the joint. Begin by tacking the joint to hold the two panels securely. Then grind the tacks level with the surface of the panels so that they do not interfere with the final weld. Run in the joint using the stitch welding facility; this is normally operated by releasing and holding the trigger on in quick succession. There will be a timer control on the machine that sets the time

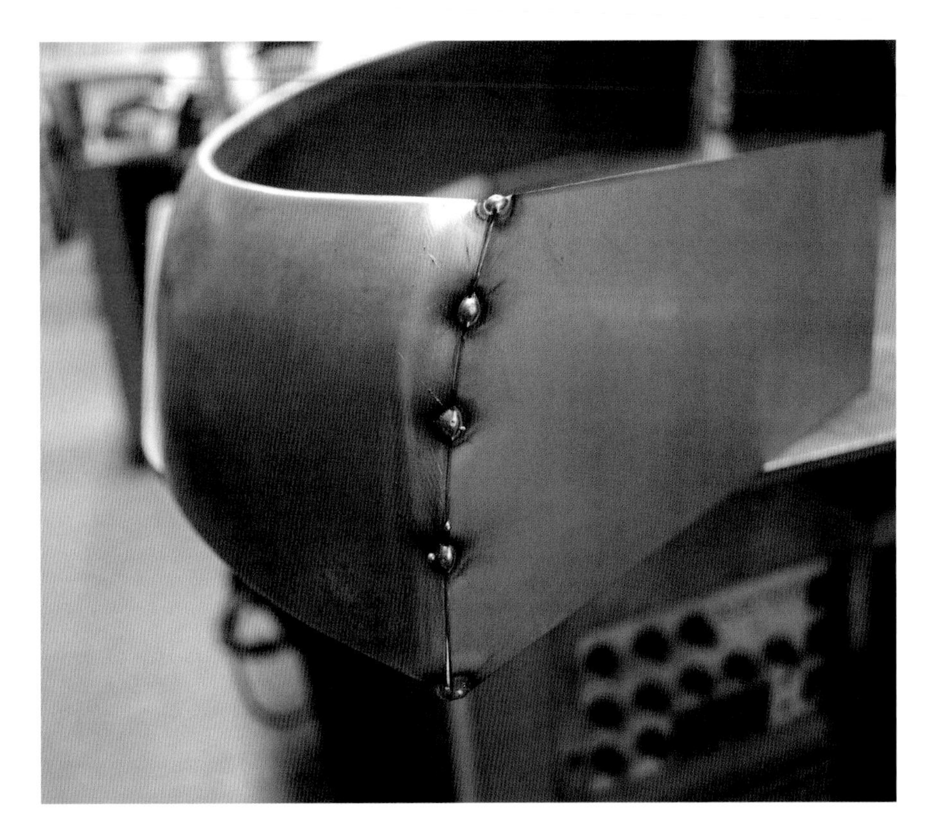

The external corner of a butt joint tack-welded with the MIG welder to hold the panels in place. Tacks are placed around 30mm apart.

The MIG tacks are ground level with the surface of the two panels prior to stitch welding the joint. This ensures a smoother finish to the stitch weld.

The stitch-welded joint. The short welds can be clearly seen. These allow for a higher amperage to be used than would be feasible with a straight run of weld.

The excess weld material is ground flush with the panel's surface. A small radius is ground on the corner. The stitch process ensures that the inside of the joint is buttressed due to good penetration, which is achieved by the high amperage used.

period for the stitch process. Keep the trigger held on and the timer will control the on and off time, giving short bursts of weld. This allows the use of a reasonably high amperage, ensuring good penetration of the joint without burning through. It will also produce a flatter weld than using a lower amperage, resulting in less work being needed to grind off the excess material.

CHAPTER SEVEN

TUNGSTEN INERT GAS – TIG

APPLICATION OF TIG WELDING

TIG welding has generally superseded oxyacetylene gas welding, which was traditionally used as a heat source for joining panels when forming a butt weld (with or without the use of filler rod) on vehicle bodywork. There are several advantages to using a TIG welder over oxyacetylene gas: minimal distortion is created due to a narrow heat spread around the weld; the process is safer, in that the gas used to shield the weld is non-combustible; and time is saved in the preparation of the joint and cleaning the joint after welding, as no flux is required for the process. The TIG torch can also be used in a similar way to oxyacetylene as a general heat source for heat-shrinking a stretched area of a panel, or heating a heavier plate or bar to facilitate bending.

Most of this chapter is focused on explaining the process of using a TIG welder without the use of a filler rod, as the aim when butt welding sheet metal is neither to increase nor decrease the thickness of

TIG butt-welding aluminium alloy double-curvature panel on a coach-built body.

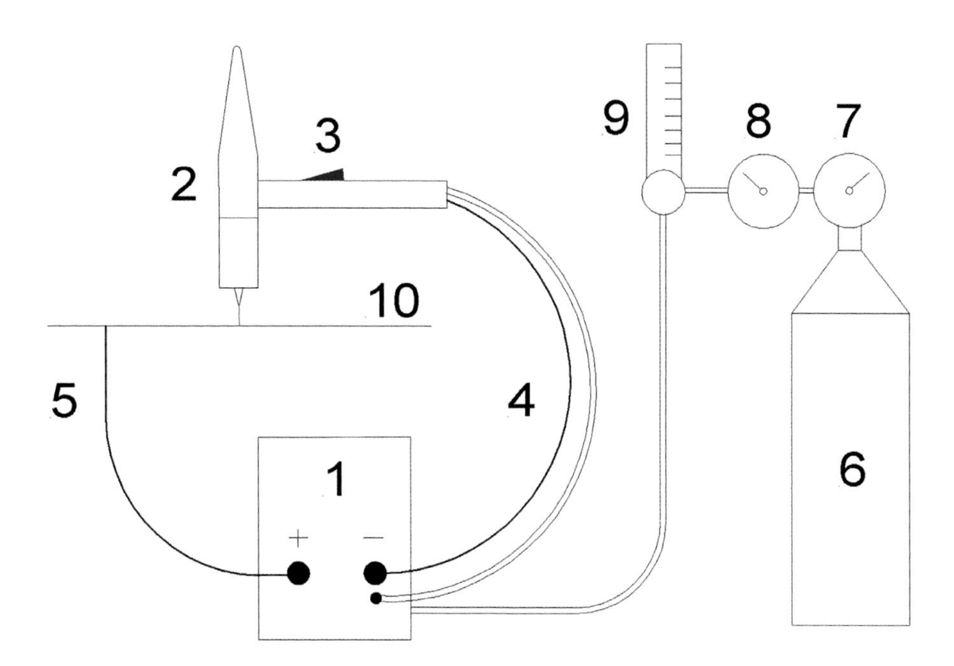

TIG welding process: 1) welding machine – provides DC or AC current supply; 2) torch – holds tungsten electrode and directs gas flow around weld; 3) switch – controls supply of power and gas to torch; 4) power and gas supply cable – supplies power to electrode and gas to torch; 5) earth lead – connects power supply back to welder; 6) gas supply – pressurized bottle containing pure argon gas; 7) gauge – shows contents of gas bottle; 8) regulator – controls pressure released from bottle; 9) flow meter – controls flow rate of gas supplied to torch; 10) workpiece – metal being welded.

the metal on the welded joint. We are simply fusing the edges of the panels together. This process is particularly useful in the repair of corroded panels, as it creates an invisible repair to a sheet-metal panel with no change in the structure of the metal on the welded joint, allowing it to be planished (hammered) to a smooth finish. There are occasionally situations that will require the use of filler rod; this is explained towards the end of the chapter.

All weldable metals can be welded using the TIG process. DC is used for most metals; AC is used for welding aluminium alloys.

THE PROCESS

The TIG welding process works by striking an electric arc between a tungsten electrode and the surface of the panel being welded. This creates enough heat within the metal to melt it at the point of the arc's interface with the panel. At the same time, a shielding gas is fed around the weld pool to avoid any inclusion of atmospheric oxygen and nitrogen into the weld. A tungsten rod is used for the electrode, as tungsten's melting point is very high (3,400°C), so will not melt during the welding process.

SETTING UP EQUIPMENT FOR 0.9–1.2MM MILD STEEL

SHIELDING GAS

The gas used for welding all metals using the TIG process is pure argon. The flow rate of the gas is set at 4–5lpm. Too much gas acts as coolant, reducing the temperature of the weld pool. Too little gas means that the weld is not sufficiently shielded from atmospheric gases, resulting in oxygen inclusions forming in the weld.

The ceramic or glass shroud fitted to the torch, which directs the gas flow to the weld joint, is a

List of equipment

- TIG welder – machine with AC facility for welding aluminium alloy; DC for welding all other metals
- argon gas – for welding all metals using the TIG process
- flow meter – for precise control of gas flow to weld pool
- torch – zirconiated tungsten fitted for welding aluminium alloy; thoriated tungsten used for all other metals
- torch holder – suitable for holding TIG torch securely to work bench of vehicle body
- TIG welding helmet – fitted with a passive lens with a number nine shade
- small-headed hammer – flat polished face for levelling welded joint
- wire brush – stainless-steel bristles for cleaning aluminium, brass or stainless steel welds; steel bristles for use on mild steel
- welding stool – with adjustable height and tilt mechanism
- gloves – heat-resistant gloves to protect hands and wrists from burns
- fire extinguisher – keep a hand fire extinguisher close by to put out any flames that may be started by sparks thrown out from the welding process.

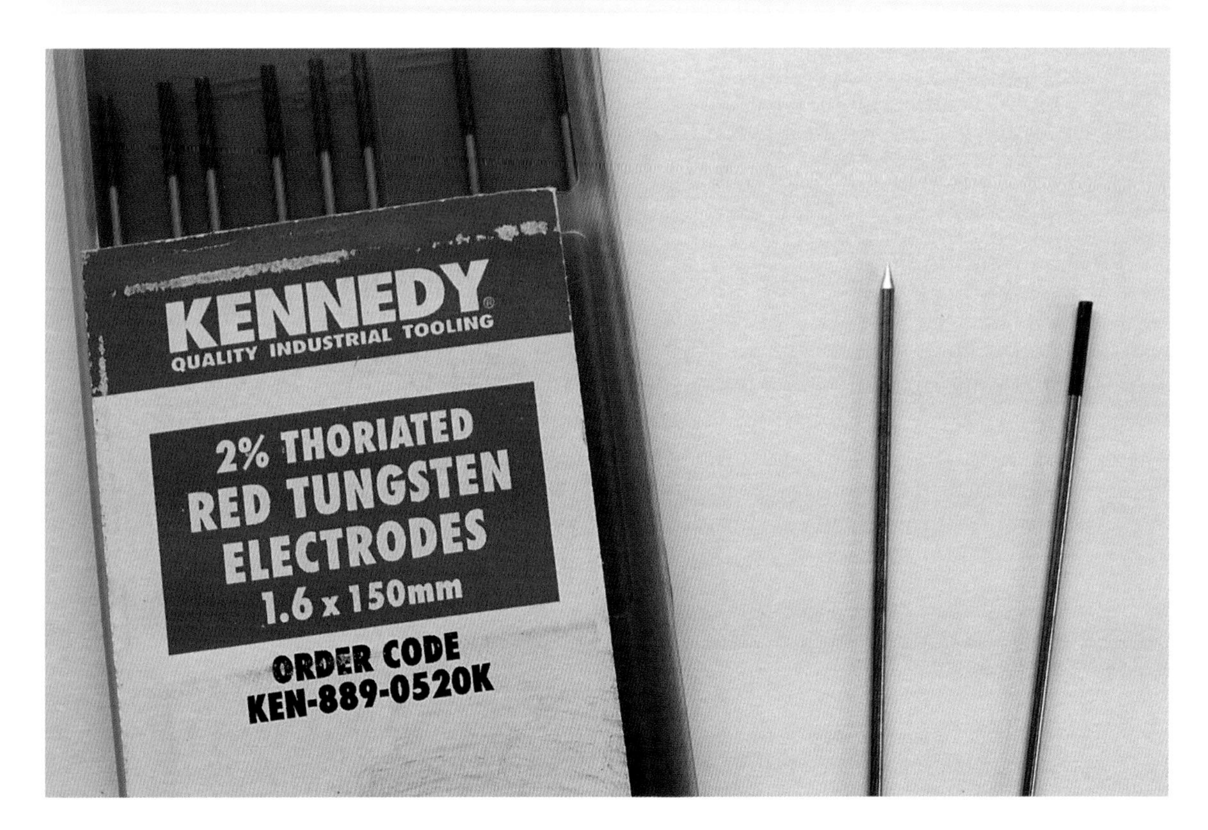

A thoriated tungsten electrode is colour-coded red and sharpened to an included angle of around 60 degrees. This must be kept clean and sharp to achieve a good weld.

number 6. Its size is related to the width of the weld joint and how far from the surface of the panel the torch is held. The welding current is set at 36–40amp DC.

ELECTRODE

Use a 1.6mm diameter thoriated tungsten electrode. This is a tungsten rod alloyed with a small amount of thorium to give it the desired characteristics for welding mild steel. The tungsten electrode carries the current to the weld joint, so its diameter will always be related to the amperage used. Too large a diameter electrode creates an unstable arc that will wander; too small a diameter electrode results in the point of the electrode melting. A thoriated tungsten is colour-coded red on the end of the shaft to distinguish it from other forms of tungsten electrode. The electrode is sharpened to an included angle of 60 degrees and must be kept sharp and clean to produce a clean, strong weld.

The tip of the electrode should protrude slightly from the end of the torch shroud; the amount that the tungsten protrudes is not critical and will generally be the sharpened point plus 1–2mm of the shaft. The position of the tungsten is adjusted by unscrewing the torch back cap that slackens off the collet holding the tungsten in place; tightening the back cap secures the tungsten in place.

BEFORE STARTING

Check that the tungsten electrode is the correct diameter for the amperage being used. Check that the tungsten is sharp and clean. Holding the torch away from the panel, press the torch trigger for two to three seconds to purge any atmospheric gas contained within the torch lead. This ensures that the shielding gas is available at the nozzle as soon as you start the welding process.

TESTING THE WELDER

Run a short weld on a test piece of scrap sheet metal to check that the weld produced is clean and has a sufficient degree of penetration to ensure that both pieces of metal are joined. The weld should be slightly raised from the surface of the panels in profile, with no visible joint line on the reverse. Check that the current being used is adequate to melt the metal effectively and that the gas is shielding the weld sufficiently. No brown discoloration should be visible around the weld and the surface of the weld should be silvery blue in colour.

The welding machine needs to be earthed to the workpiece to achieve an electrical circuit to create a weld. Clean all paint from the surface of the panel on and immediately around the area, so that the earth clamp is closely fitted to ensure a good earth connection. Always test the welder on a scrap piece of sheet metal to ensure that everything is working correctly before attempting to tack or weld on body panels.

PREPARING THE JOINT

To achieve a smooth weld, it is essential that the joint is cut accurately to provide a tight butt joint with no gaps. No filler rod is used for this method as the addition of extra metal into the joint increases the difficulty of achieving a satisfactory finish to the weld. Any extra metal added alters the thickness of the material unequally on the joint, making it difficult to planish to a smooth finish. Check that the metal is clean and free from rust, oxide, paint or oil contamination. **When making a butt joint, it is essential that the panels are cut accurately to provide a tight joint with no gaps.**

WELDER OPERATION

A clear view of the weld pool is critical; use reading glasses if necessary, to enable you to see clearly at a short distance. Avoid glasses fitted with bifocal or varifocal lenses, as you will need to tilt your head as you progress, meaning that your work may go out of focus. A work light aimed at the joint can also help to provide a clear view of the weld pool. Lean your upper body forward and turn your head sideways, looking along the joint so that you can clearly see whether the joint is pulled tight together and that both panels are level with each other.

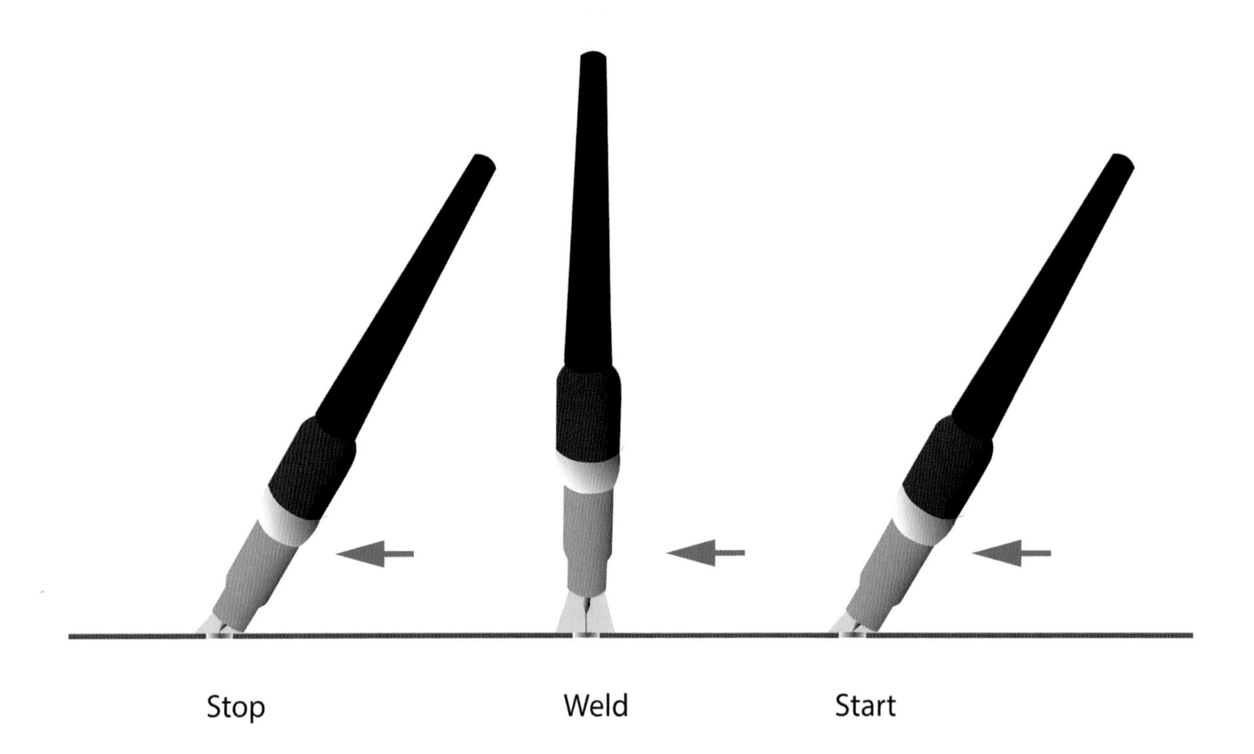

Hold the torch shroud on the panel at the start and end of the weld to avoid lifting the shielding gas away from the weld pool. If the torch is lifted too high from the weld pool it will create an inclusion of atmospheric gases into the weld.

Take care to avoid burning the metal. This is caused by inadequate shielding of the weld due to too low a gas flow rate, or holding the torch too high from the surface of the panel, or too much power being used for the thickness of the material being welded.

When reaching the end of the weld, the natural tendency is to lift the torch away from the panel when releasing the trigger. **It is critical to avoid this,** as it results in an inclusion of atmospheric gases into the weld pool because the shielding gas has been removed while the metal is still molten. This is a very common fault and practice is needed to train yourself to lift your finger off the torch trigger without lifting the torch away from the weld pool. Most welding machines will automatically have a post-flow rate of one second. This is the period for which the gas will continue to flow after the weld current is stopped, helping to shield the molten weld pool from the atmospheric gases as the weld cools. The post-flow rate can be set for a longer period when welding thicker materials; one second is adequate when welding thin sheet metal.

Practise starting and stopping the arc on a scrap piece of sheet metal to become proficient at this procedure before welding on a vehicle body panel. A useful technique to practise, particularly for the beginner, is to start the arc with the edge of the shroud contacting the panel and with the torch at a slight angle, ensuring that the tungsten is not in contact with the surface of the panel. Once the arc is running, you can lift the torch away from the surface of the panel and hold it near vertically to the surface of the panel, to continue welding.

When stopping, the reverse procedure can be used. Place the edge of the shroud on the surface of the panel at an angle to avoid contacting the electrode to the surface of the panel, at the same time as letting go of the trigger to stop the arc. This ensures that the gas is still shielding the weld pool as the metal cools. A good indicator of this is the colour of the finished weld: silver grey is ideal; silver blue is acceptable. Any brown discoloration around a weld is a sign that the weld pool has not been adequately shielded by the gas.

TACKING THE JOINT

Clamp both panels together at one end only and work from the clamped end, placing tacks around 15mm apart. This differs from welding heavier plate or sections that would be tacked at both ends first. Sheet metal will distort as you build heat into a panel, so it needs to be tacked progressively along the joint to maintain a level and tight joint. The technique is similar to closing a zip, in that as you work along the panel it will pull in, keeping the joint tight and level as you progress. Use your non-torch hand to control the level and tightness of the joint at the position you place each tack.

ELECTRODE

The electrode is positioned around 3–4mm from the surface of the panel. A tack weld is achieved by first heating either side of the joint to expand the metal; this ensures that the joint is tight where the tack is placed. Heat the area adjacent to the tack, then place the welding arc directly over the joint and hold the torch still until a small molten pool of metal is formed. Release the trigger as soon as the pool forms to create a small area of weld between the two panels. **Keep the torch close to the weld pool after releasing the torch trigger to shield the weld as it cools.**

The natural tendency is to lift the torch away from the panel when releasing the trigger, but it is crucial to avoid this. Lift your finger off the trigger without

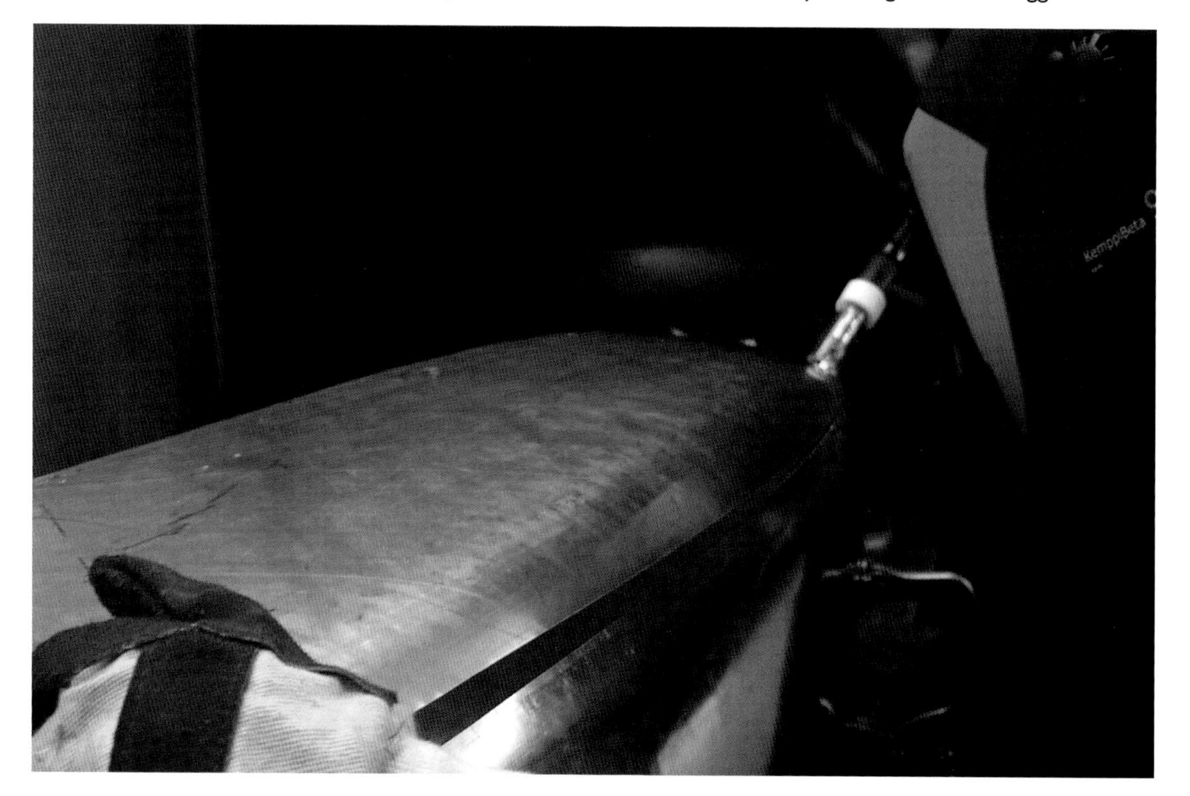

Clamp the two panels together at the end of the joint where you will place the first tack. Use your non-torch hand to hold the panels level at the position of each tack as you progress along the joint.

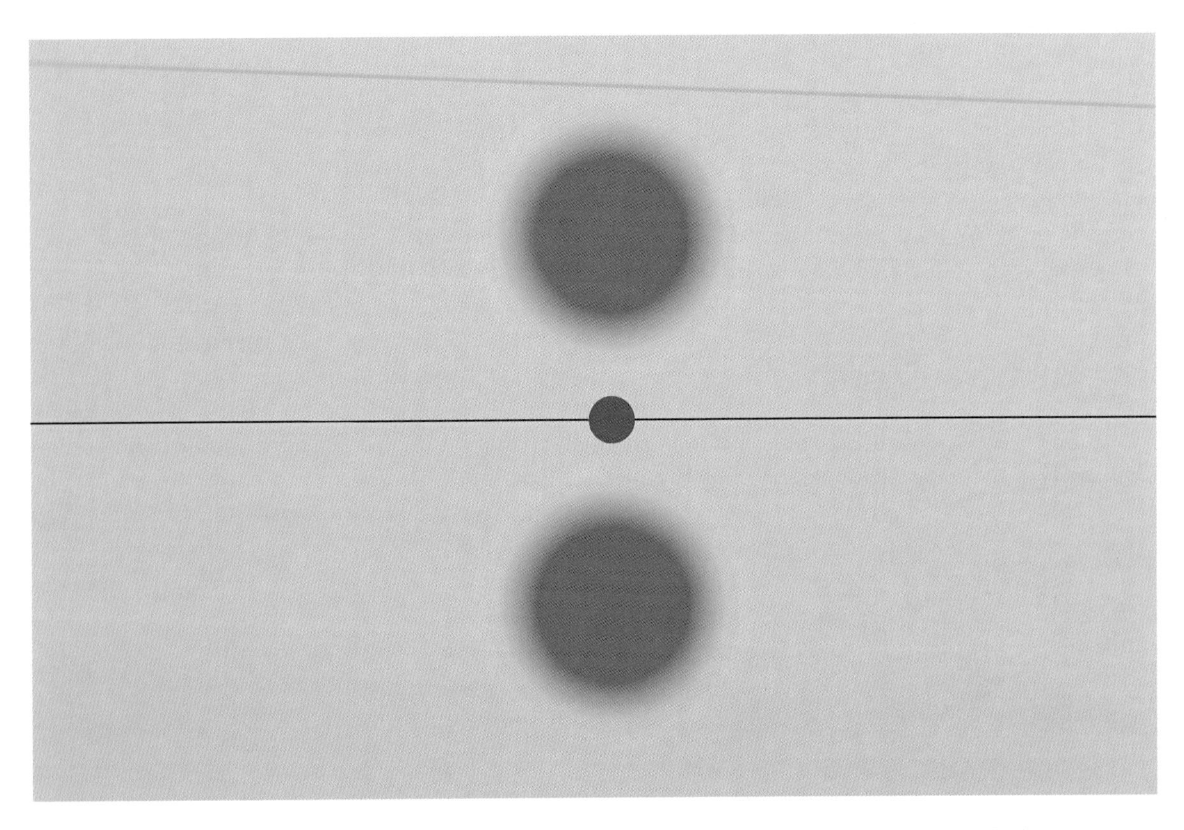

Heat an area adjacent to the joint on both sides of it so that the metal expands to push the two panels tight together at the position of each tack.

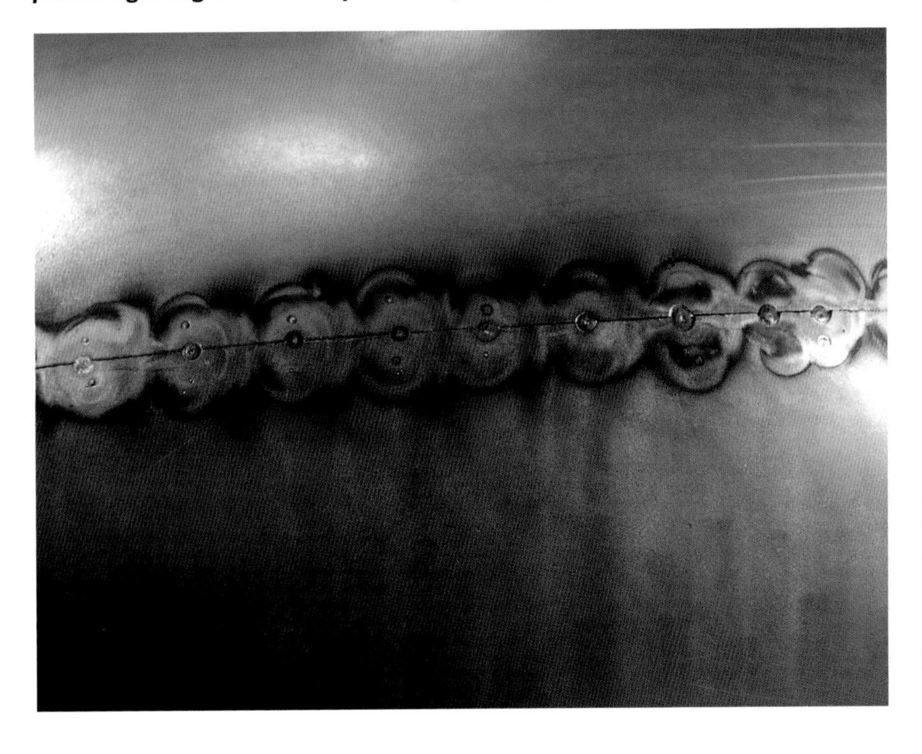

Tack welds should be around 2mm in diameter and silver blue in appearance. Place at 20mm intervals to hold the joint securely prior to running-in the weld.

Remove the surface oxide created by the tack welds using a nylon sanding disc that cleans the surface of the panel without reducing the thickness of the metal.

lifting the torch away from the weld pool. Failure to do this will result in an inclusion of oxygen into the weld pool as the shielding of the gas is removed while the metal is still molten. As the weld cools and the metal shrinks the joint will close. Patience is required for this process; do not attempt to create further tacks until you are confident that the joint of a tack is tight and level. If the tacked joint fails to close after welding, run in a short length of weld between the last two tacks. This will help pull to the joint tight as the welded section cools and shrinks. If the joint pulls in too tight, creating an overlap of the panels, this can be resolved by hammering on the last tack to relieve the tension in the panel and allow it to be pulled level on the joint.

Tacks should be kept small, for example 2mm in diameter on 1mm thick steel. A larger tack will create an inclusion (air bubble) on the rear of the weld due to a lack of shielding gas on the back of the joint. Ensure that the panel is tacked adequately to hold the joint tight and level along the complete length of the joint.

Descale the tacks before running-in the weld using a nylon wire wool pad or nylon sanding disc that will remove any surface oxide without grinding away the metal. Lightly hammer the tacks against a dolly held behind the joint to level both panels before running-in the joint to create the final weld.

Position your head to the side of the weld so that you are looking along the joint as you move the welding torch towards you. It is crucial that you can see the weld pool clearly as you progress, so move your head at the same speed as the torch.

RUNNING-IN THE JOINT

TORCH CONTROL

Controlling both the speed of traverse and distance of the tip of the torch nozzle from the panel is critical to ensuring a consistent, smooth finish to the weld. To maintain a steady movement, rest the torch hand in your other hand, moving your upper body with the torch to maintain a constant focal distance from the weld pool. For a right-handed person, position yourself to the left of the torch, looking along the joint towards the torch; for a left-handed person, position yourself to the right of the torch, looking along the joint towards the torch. Lean the torch away from you at a slight angle to give a clear view of the weld pool.

Leaning the torch at too great an angle from 90 degrees to the face of the panel will mean that the gas is not shielding the weld pool adequately; make sure that this is no more than 10 degrees from the vertical. Be aware that the arc will always take the shortest path to earth, so however you angle the torch the arc will not vary, but the direction of the shielding gas will.

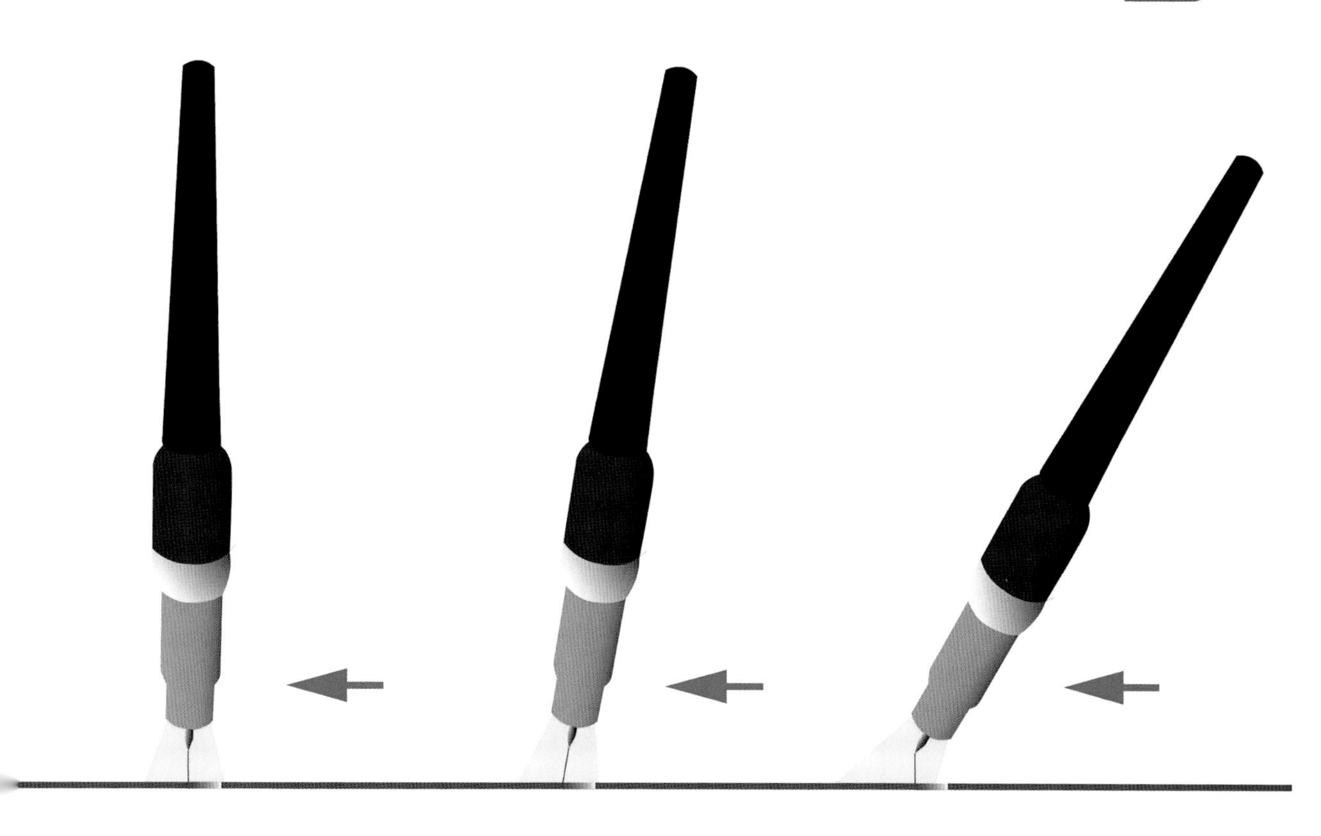

Maintaining a reasonably upright angle to the torch during TIG welding is important in order to avoid a lack of shielding gas covering the weld pool as it cools.

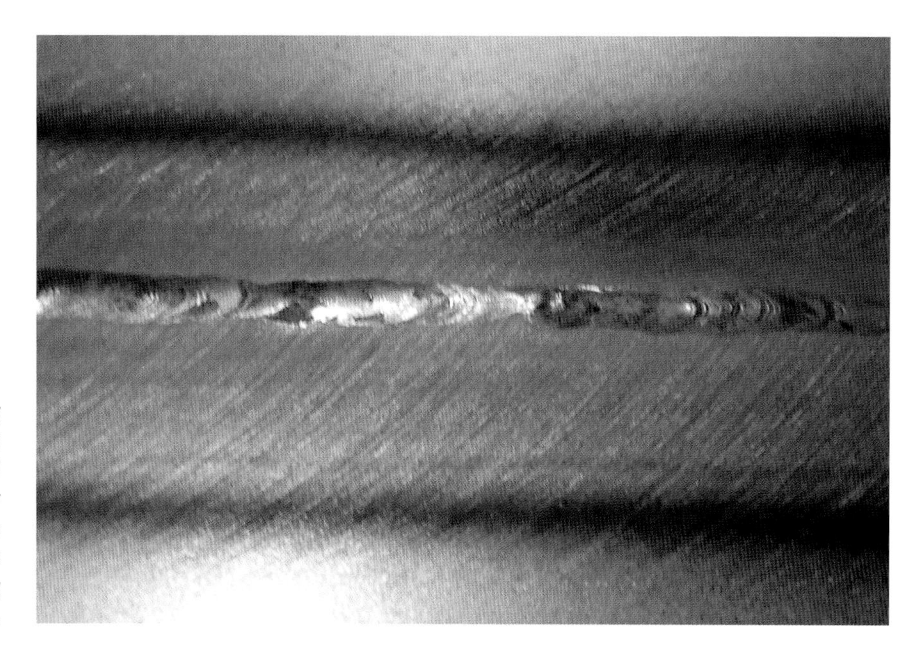

Focus on attaining a consistent width of weld. The speed of traverse will vary depending on the tightness of the joint at each point along the seam.

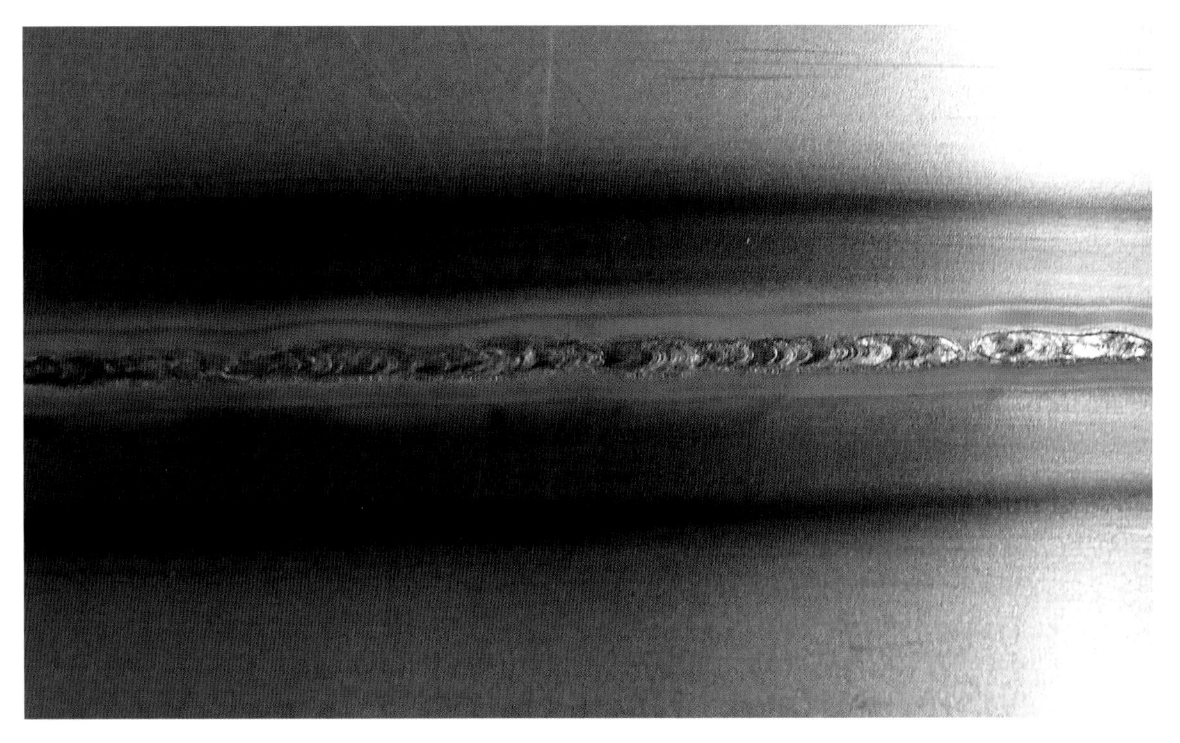

The back of the weld will not be as clean as the front due to the lack of a shielding gas covering this side of the weld joint.

Always move the torch towards you to give a clear view of the joint as you are fusing it together. Use a welding stool if possible, to steady your body as you weld; any instability in your stance will create an erratic weld.

WELD WIDTH

The aim is to maintain a consistent width of weld. The speed of traverse will vary depending on the tightness of the joint and will need to be sped up as more heat builds into the panel. The width of the weld needs to be around 3mm in order to achieve enough penetration and to avoid the weld dipping on the surface. Ensure that you have a clear view of the molten weld pool so as to maintain a constant width of the pool as you move along the joint. Too narrow a weld means that the traverse speed is too quick; too wide a weld means that the traverse speed is too slow.

Inconsistency of the width of the weld is due to poor control over the traverse speed. This could be due to an uncomfortable stance or lack of hand control. Make sure that you are sitting comfortably if possible and use both hands to control the movement of the welding torch. The surface of the weld, if done correctly, will be slightly raised. This is caused by shrinkage of the metal across the width of the weld pool as it cools. Vary the amperage slightly to suit the speed of welding that you find comfortable, whilst still achieving the correct penetration of fusion on the joint.

There should be no visible line on the reverse of the joint – this would indicate a lack of sufficient penetration of the weld. A significant amount of practice is required to attain the necessary speed of traverse to achieve the correct width of weld and level of penetration.

There will be a certain amperage that works best on any given thickness of metal to achieve the correct finish to the weld. Practise on test pieces of sheet metal using different amperage settings to discover what works best on your machine for the thickness being welded.

The width of the blue area of heat spread around the weld is a good indicator of the consistency attained during the welding process. This should be parallel.

MINIMIZING AND RELIEVING HEAT DISTORTION

The distortion of panels from welding is caused by heat shrinkage on and around the joint. As soon as heat is applied to metal it expands; it is then joined by welding whilst in the expanded state. As it cools, the joint and surrounding metal shrink, causing distortion as the panel is now pulled into tension. The distortion is minimized by keeping the weld as narrow as possible, then by welding as quickly as possible, so that there is minimal time for heat to spread into the surrounding panel.

A constant width of weld will cause less distortion than an inconsistent weld. Aim to keep the area of blue heat spread the same width over the entire length of the weld.

Distortion is relieved by hammering on the weld using a small-headed hammer with a steel dolly held directly behind the area being hammered. This flattens and stretches the metal, relieving the distortion. Do this in as consistent a manner as possible in order to achieve a smooth finish. It is best done when the weld is still hot. Allow the metal to cool fully before deciding whether the joint requires further hammering, as the panel will pull flatter as the metal cools.

SETTING UP EQUIPMENT FOR 1.2–1.6MM ALUMINIUM ALLOY

The gas used is pure argon and the flow rate is set at 8lpm. The ceramic shroud used is number 8. The welding current is set at around 80amp AC to tack the joint. This is reduced to around 60amp to run-in a weld.

A common mistake made when learning to weld aluminium is lowering the amperage to avoid blowing holes. The opposite happens in practice, as the metal will not flow across the joint if too little amperage is used, which then causes a hole to open around the joint. Set AC frequency at 100–120Hz, AC balance at 50 per cent.

A zirconiated tungsten electrode is colour-coded white. This naturally forms a ball on the end as it is used.

ELECTRODE

The electrode used is a 2.4mm diameter zirconiated tungsten (tungsten alloyed with a small amount of zirconium to give it the desired characteristics for welding aluminium alloy). A zirconiated tungsten is colour-coded white on the end of the shaft.

The electrode will naturally form a ball on its end as it is used, which is preferential as it creates a wider electric arc so producing a wider weld. The electrode must be kept clean to maintain a reliable arc. The electrode is carrying the current to the weld joint so it's diameter will always be related to the amperage used; too large a diameter electrode creates an unstable arc that will wander around. Too small a diameter of electrode results in the end of the tungsten melting. A zirconiated tungsten electrode can develop a small crack along its length near to the balled end or craters in the balled end, both defects will cause an unstable arc and give a poor weld. If a crack is evident grind the end of the tungsten well back from the end of the crack to avoid it re-occurring. If the end is pitted or cratered, grind the end back to clean metal. Reform the balled end by running a weld along the surface of a scrap piece of sheet metal at a high amperage. It is critical that the tungsten is kept clean! Avoid touching the end of the tungsten which will contaminate it.

Defects in the zirconiated tungsten: the left-hand one shows a crack along the length of the shaft; the centre one is pitted on the end of the shaft; the right-hand tungsten is correct.

TACKING THE JOINT

To achieve a smooth weld, it is essential that the joint is cut accurately to provide a tight butt joint with no gaps. No filler rod is used for this method. The metal needs to be clean and free from any contamination. The electrode is positioned around 5–6 mm from the surface of the panel. A tack weld is achieved by first heating either side of the joint to expand the metal; this ensures that the joint is tight where the tack is placed.

The method used for aluminium alloy differs from welding steel. Heat one side of the joint with a short blast of the arc, release the torch trigger to break the arc, then immediately start an arc on the opposite side of the joint. Keep the arc going on this second spot and run a molten pool of metal on to and along the joint for a short distance (around 6–8mm). Run off the joint as you release the trigger to avoid blowing a hole at the end of the tack. Avoid lifting the torch away from the surface of the panel as you lift your finger off the trigger. This is a common

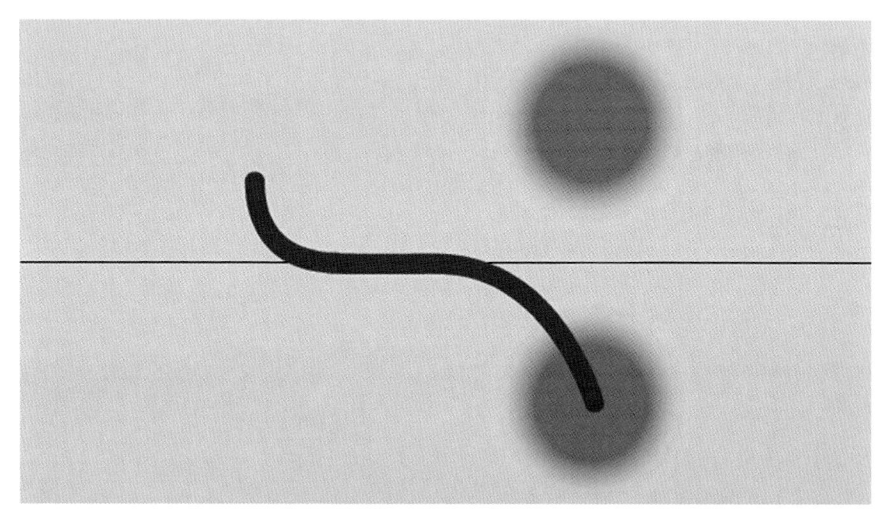

When welding aluminium, the tacking process is carried out without stopping at the start and end of the tack to avoid blowing holes at either end of the weld.

Tack welds on aluminium alloy panel showing the 'S' shape of the tack. These are spaced around 25mm apart.

fault, as the natural reaction when stopping the arc is to lift your hand away. Practise on a scrap piece of metal, stopping the arc without lifting the torch until this becomes an instilled habit.

The width of the tacks should be kept narrow – around 3mm wide. A larger tack is likely to create a small hole at either end, which may result in blowing through the panel as the weld is run-in. The tacks should be placed around 25mm apart. Ensure that the panel is tacked adequately to hold the joint tight and level. It is not necessary to descale tacks, as the action of the AC arc removes the surface oxide during welding. Lightly hammer the tacks to level both panels before running-in the joint. Hammering too forcefully on the tacks will open up the joint, which is to be avoided. Be vigilant in keeping the metal clean, as any contamination will cause problems while the joint is being run-in. An arc that is non-directional and wanders signifies contamination of the metal being welded or of the electrode, or damage to the end of the electrode.

RUNNING-IN THE JOINT

Use the same method as on steel. Start a run of weld slightly off the joint just before a tack to avoid blowing a hole. The width of weld needs to be around 4mm to achieve sufficient penetration to avoid the weld dipping on the surface. The surface of the weld should be left slightly raised. This is caused by shrinkage across the width of the weld pool as the metal cools.

Too high an amperage used or too slow a traverse speed will result in the weld dipping on its surface. This is to be avoided, as it will create a thinner area of metal on the weld and leave a visible edge to either side of the weld following any finishing.

Check the back face of the weld for sufficient penetration. No joint line should be visible, as this shows that the weld has not penetrated sufficiently. There will be an optimum amperage and traverse speed to match this for the thickness of the metal being welded. You will need to experiment using various

The completed weld should be around 4mm wide and left slightly raised in profile from the surface of the panel.

A dip in the weld's surface is created by using too high an amperage or too slow a traverse speed.

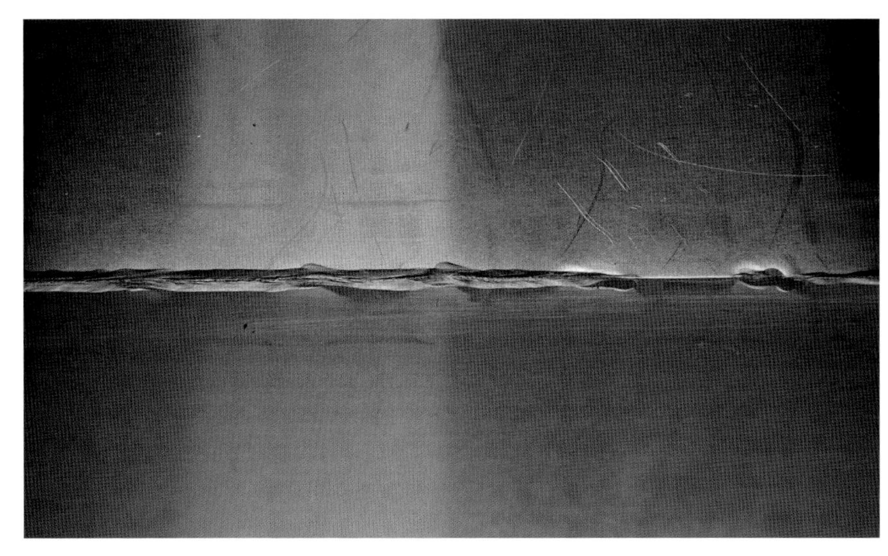

The back face of the weld: the right-hand side shows a lack of penetration where the line of the joint is visible; the left-hand side shows the correct level of penetration.

amperage settings to determine what works best for the material.

You will occasionally get a gap in the joint. This may be caused by a poor tack weld or a poorly cut joint. Any visible gaps left following the tacking process should be avoided when running-in the weld. It is best to mark these areas with a permanent marker pen so that they are clearly seen as you run-in the weld.

Gaps or holes can usually be filled by hammering the metal on either side of the gap. Use a small-headed hammer to ensure accuracy with a steel double-curvature dolly held behind the panel. This will close the gap and can then be run over with the torch to melt the metal, so filling the gap. Make sure that there is no visible hole, by holding the joint up to a light source, before attempting to weld over the hammered area. Any holes that cannot be filled by this process can be repaired as described in Chapter Nine.

A weld on aluminium alloy may be insecure due to an internal crack that is caused by the metal shrinking

A gap in the joint resulting from a poor tack weld that has melted the edge of the panel back from the joint.

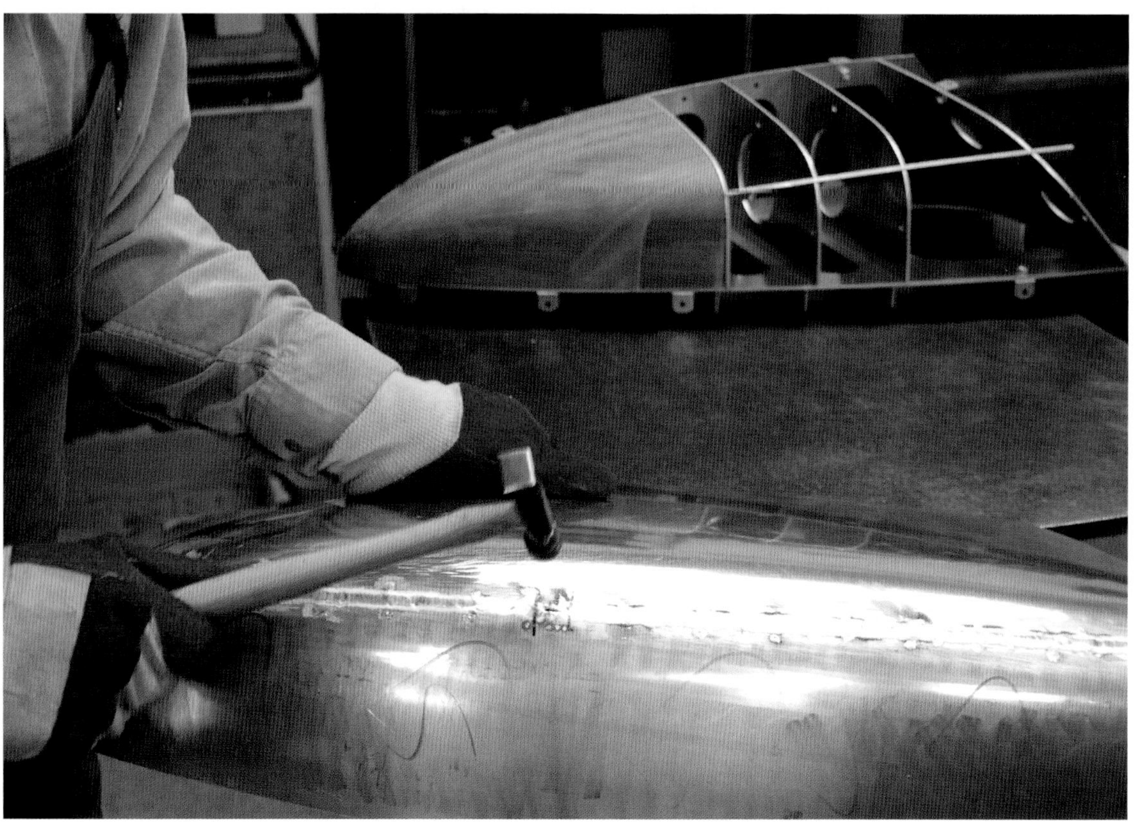

Hammer on either side of the gap, which compresses the metal and spreads it into the gap. Run back over the joint with the metal together.

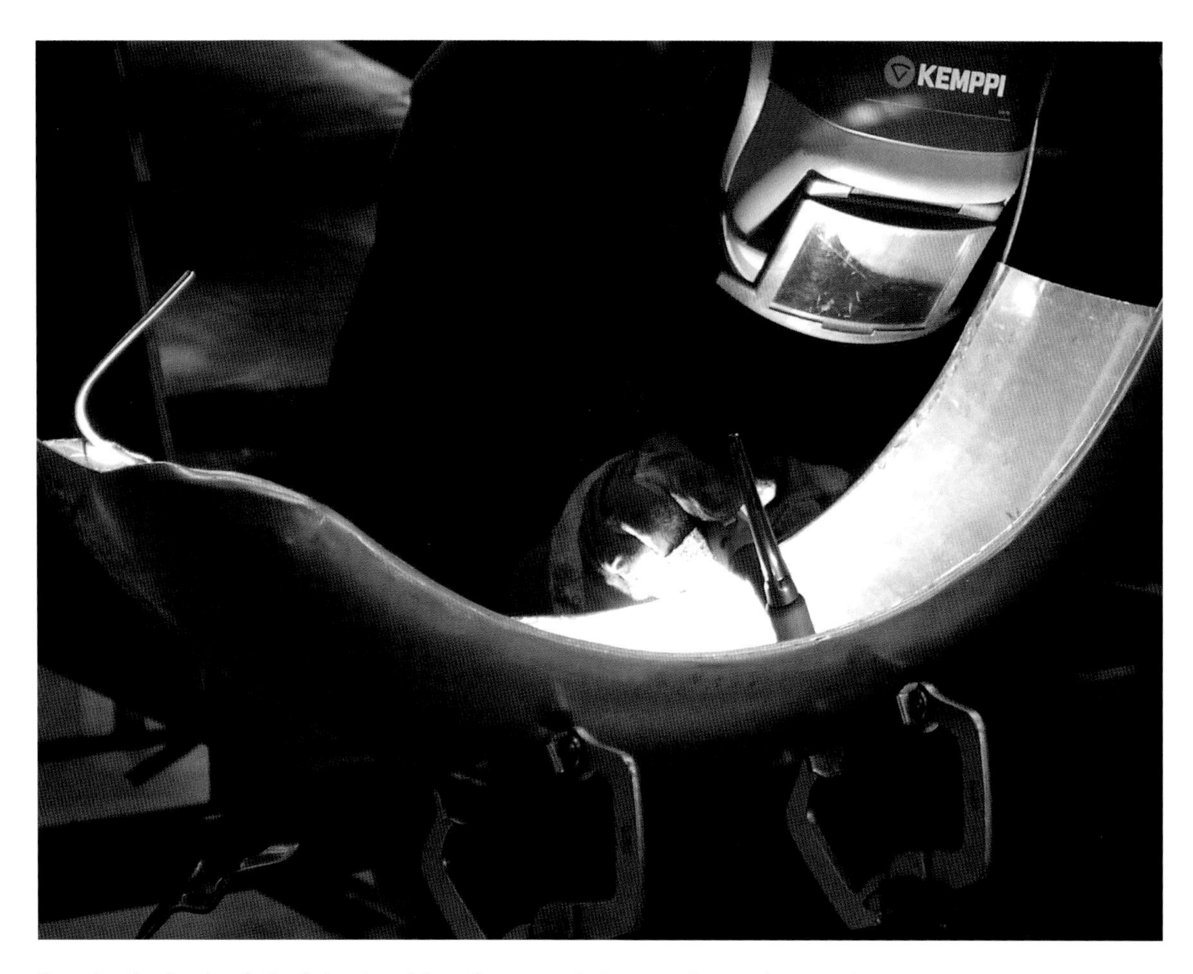

Run-in the back of the joint to widen the annealed area of metal around the weld and avoid a stress crack occurring within the weld.

away from the joint as the weld cools. This will not be visible on the surface of the panel but can cause a weld to fail and crack if any stress is placed on it. To overcome this problem, it is necessary to weld along the joint more than once – this creates a wider annealed area around the weld, which acts as a buffer as the metal shrinks, so avoiding an internal crack in the weld. This can be done on the same side, or on the back face of the joint. Aim for as wide a weld as possible to also help alleviate this problem. It is best to weld over the joint three times to ensure that a robust weld is made without the risk of any fracture of the joint when placed under stress.

The weld is finished in a similar way to steel. Aluminium alloy will be annealed on the welded joint due to the heat used in the welding process, which makes the metal significantly softer than the surrounding panel. The weld and surrounding area will need to be hardened by hammering or wheeling, which also levels and smooths the joint at the same time. Finish by sanding using a coarse fleece flap disc to remove the minimal amount of metal and leave a smooth finish.

USING FILLER ROD

A filler rod is used on a buttress or overlapped joint, usually when welding a heavier gauge bracket to a chassis or main body member. It is also used when tacking a panel to a frame, such as on a door, bonnet or boot lid. The rod should be of the same material and thickness as the panels being welded. This can be

Using a filler rod to fill an open corner joint. Keep the torch square to the corner of the joint being welded to ensure that the gas flow is shielding the weld as extensively as possible.

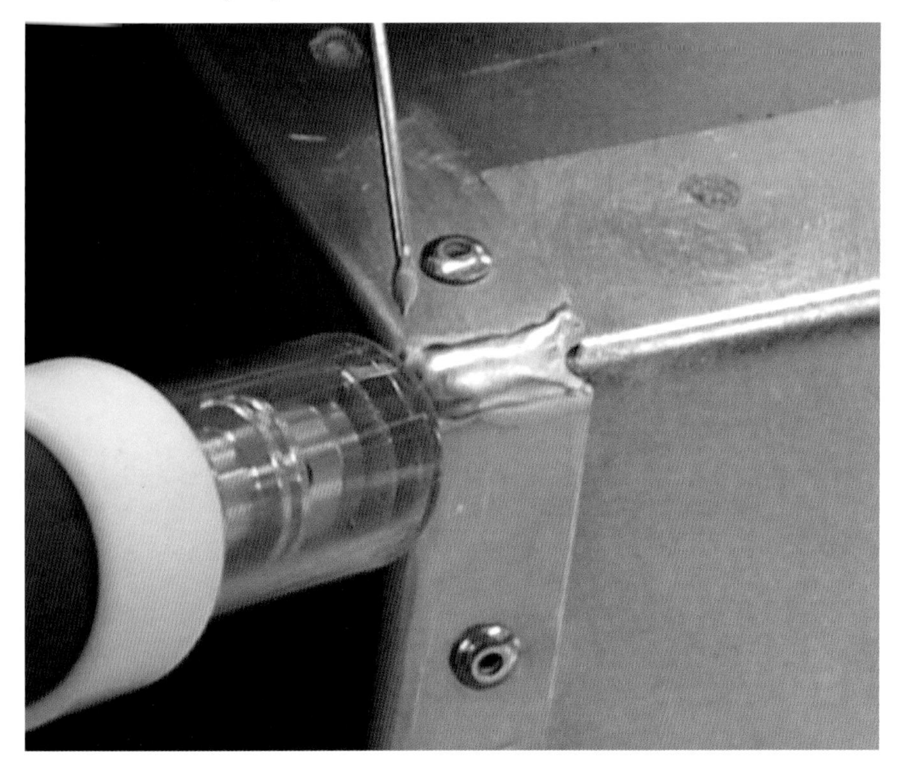

The filler rod is dipped into the weld pool intermittently to avoid burning the metal on the end of the rod.

cut as a thin strip from the edge of a piece of a metal sheet, but be aware of contamination from the blades of the cutting tool that is used.

The rod must be fed into the weld in an intermittent manner. If it is fed continuously it will tend to burn and create problems due to contamination of the weld pool. The rod is best fed into the weld at a shallow angle (10–20 degrees). Panels being tacked are first heated with the torch to create a molten pool. The rod is then dipped into the leading edge of the weld pool. As the rod is retracted, a lump of metal is deposited and then melted into the surface of the panels being joined. This all happens within a short time and a significant amount of practice is needed to produce a clean, effective tack while avoiding over-melting the panel or burning the rod.

SPOT WELDING

The TIG machine can be used to replicate an electrical resistance spot weld. To achieve this, the tungsten electrode is pushed further back into the collet, setting it back 2–3mm from the edge of the shroud. The shroud is placed flat on the surface of the panel. The machine is set to a high amperage (around 100amp) and the trigger pressed for the minimum duration necessary (one to two seconds), so that the weld penetrates both panels being joined. Practice will be needed to determine the correct amperage and time needed to produce an effective weld that melts through both panels without causing excessive distortion from the heat produced.

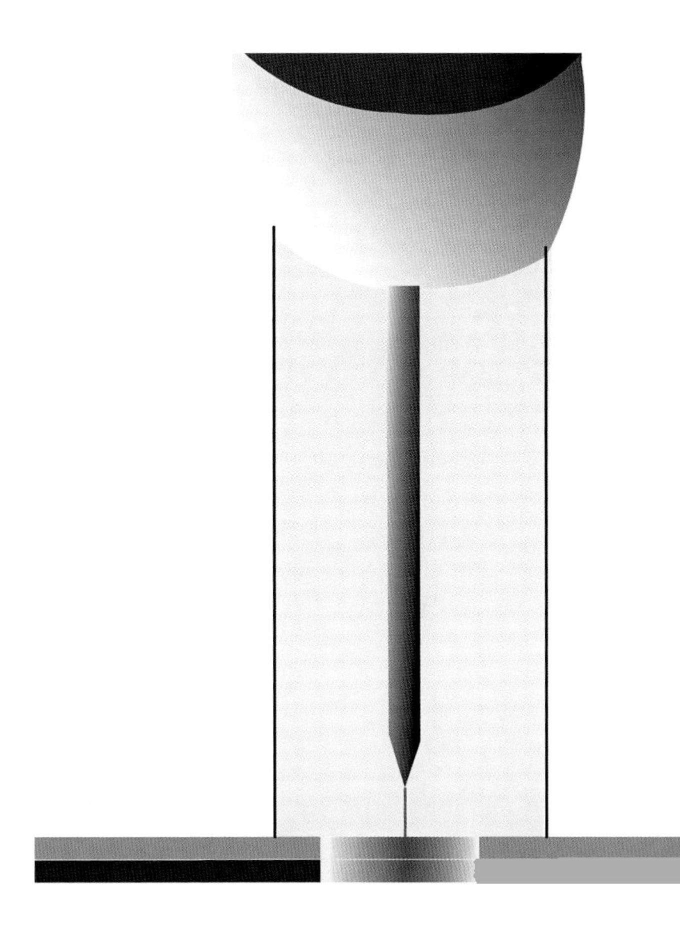

A TIG plug weld is achieved by holding the torch gas shroud flat on the panel, with the tungsten electrode set back into the torch to leave a gap between the end of the electrode and the panel. Use a high amperage with a short duration to avoid heat spread into the panels.

TIG WELDING STAINLESS STEEL

The TIG welding of stainless steel is similar in principle to welding mild steel. Pure argon gas is used for shielding the weld pool. Stainless steel flows readily when molten, requiring less heat input than mild steel to melt the metal due to its comparatively poor conductivity.

A problem experienced when welding stainless steel arises from the scale formed on the rear of the weld. This is chromium oxide, which is extremely tough and difficult to remove even with a handheld angle grinder. To avoid the build-up of scale, make sure that the weld does not penetrate through the joint more than is necessary. A flux can be used, applied to the back of the joint prior to welding. This prevents atmospheric oxygen being introduced into the weld pool. Another tactic to prevent the oxide forming is to back the joint with a strip of metal to prevent the atmospheric gases reaching the weld pool.

CHAPTER EIGHT

FITTING PANELS

PREPARING A BODYSHELL FOR RESTORATION WORK

All trim, glass, rubber seals and so on should be removed before carrying out any restoration work on the bodyshell of a vehicle. Be aware that spatter, in the form of molten metal, can be thrown out during the welding process. This can damage glass, rubber, chrome and paint. If it is impractical to remove these vulnerable parts, protect them from any weld spatter by covering them with strong tape or a metal shield.

Disconnect the vehicle battery before carrying out any welding, as damage may be caused to the vehicle's electrical circuits and components from the high current produced during welding. The welding process also produces a strong magnetic field that can damage delicate instruments.

FITTING REPLACEMENT PANELS

It is the aim when carrying out any vehicle body panel work to achieve the best possible standard of fit and finish of the panels prior to fitting them to the bodyshell. This is to minimize any grinding or filling required prior to painting the body. The joints around the doors, bonnet and boot lid shut gaps should be adjusted on the panels before fitting to the bodyshell, to achieve the desired gaps.

This usually involves the reworking of flanged edges to any repair panel being fitted. Do not expect replacement panels to be an accurate fit to the original bodywork. Even an original manufacturer's spares often require adjustment to fit your vehicle body accurately. A common practice during the assembly of vehicle bodies in the factory was for poorly fitting panels to be put to one side; these were later sold as original manufacturers' spare parts.

BODY LINES AND SHUT GAPS

Problems with the fit of any panel should be resolved early on in the fitting process, rather than leaving to the paint stage. Errors in the fit of panels will tend to cause problems further on in the restoration process if not resolved as trim panels and so on are fitted to the bodyshell. Any bodylines and door, bonnet and boot lid shut gaps need to be consistent with those of the original bodyshell. These vary on different makes and models of vehicle from 3mm ($^1/_8$in) to 6mm ($^1/_4$in).

List of equipment

- body alignment jig – necessary if carrying out major repairs or panel replacement
- welding machine – MIG, TIG or spot welder depending on type of joint being welded
- clamps – G clamps and locking pliers suitable for clamping joint securely
- torch holder – suitable for holding welding torch securely
- welding helmet – suitable for MIG or TIG depending on which process is used
- face shield – for a spot-welding machine
- small-headed hammer – for levelling tack welds
- large flat-faced hammer – for planishing joints smooth
- dollies – for hammering against
- wire brush – stainless-steel bristles for cleaning aluminium, brass or stainless steel welds
- welding stool – with adjustable height and tilt mechanism.

Gaps should be correct and altered on the edge of the replacement panel if necessary, before fitting a panel to the body.

MAINTAINING INTEGRITY OF STRUCTURES; MEASUREMENT AND RIGIDITY

The strength of welded joints and the dimensional accuracy of assembly are both critical to the integrity of the vehicle structure. There are three checks used when measuring a structure:

- parallelism: ensuring that the panels lie parallel, or at the correct taper if required
- squareness: ensuring that the structure is square, or at the correct angle of lean
- twist or wind: checking that the structure is sitting flat or level, or at the correct degree of twist.

All three checks must be correct at the same time before any structure is made rigid by welding, as once it is fully welded any alteration of one aspect will result in the other two measurements changing.

These three checks apply whether assembling or repairing the simplest of frames, or the most complicated of structures. Take measurements from original panels and structures to check against as you are fitting or repairing new panels. The doors of a vehicle bodyshell are particularly important in terms of making sure these are the same geometry as the original part. If these are not restored accurately it will be impossible to achieve consistent shut gaps and the correct alignment with the rest of the body. Patterns should be taken of the front and back end profiles of doors from the original part prior to any restoration. Mark on these patterns the points at which the measurements are taken to check the parallelism and squareness of the structure. The patterns and measurements can then be used

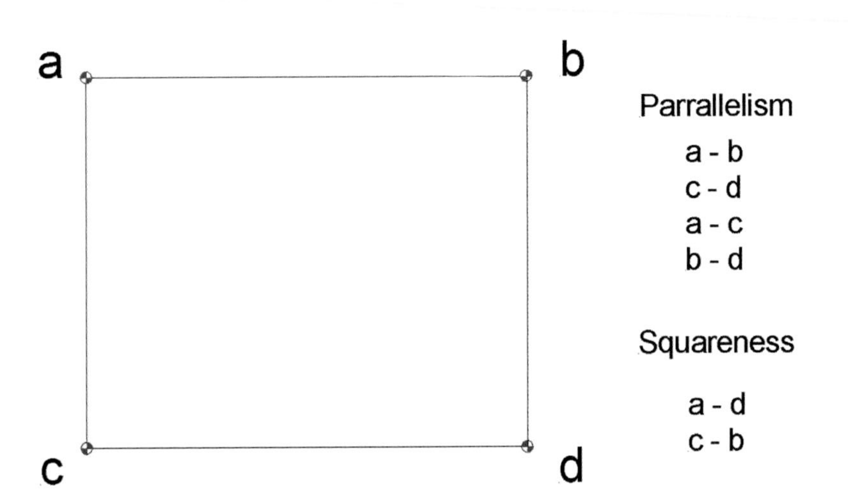

Check the dimensional integrity of a structure by measuring corner points to check for parallelism and squareness. The measurements between the points for parallelism should be identical if the structure is intended to be parallel. The two measurements across the corners for squareness should be identical if the structure is intended to be square.

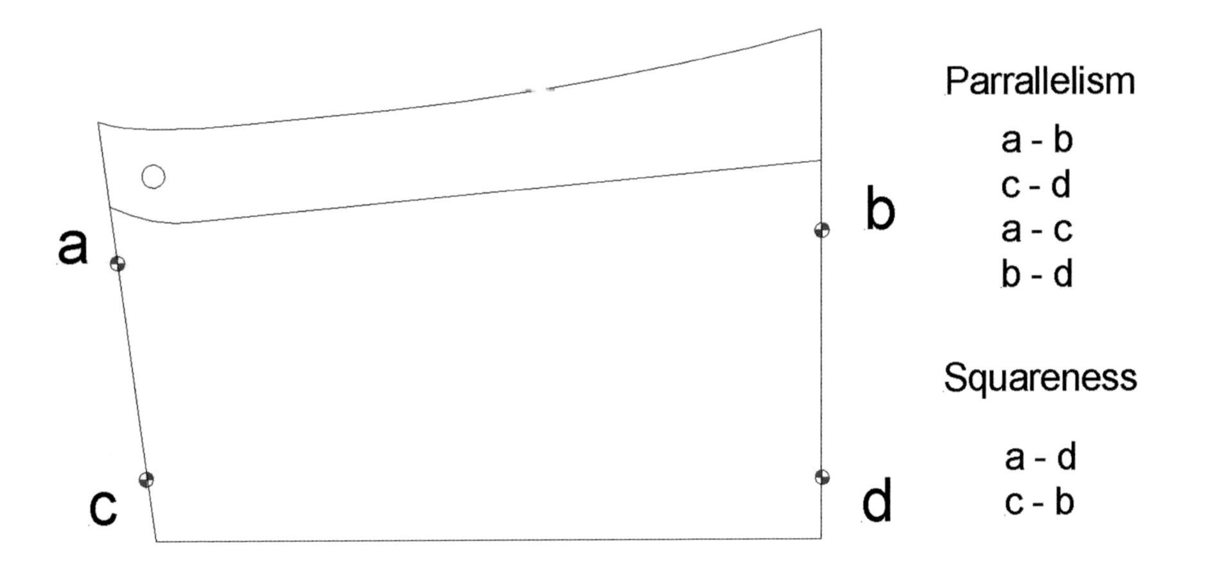

On a more complex-shaped structure, such as a door, the measuring points are placed at convenient positions marked on the panel. Measurements are taken from an original panel and then checked on the replacement or restored structure to ensure that it is dimensionally correct.

Check flatness of structure by sitting panel on a flat surface or sighting across the panel using winding sticks. Winding sticks must be sitting parallel to each other to give a true sighting.

Checking that the headlamp pods are aligned with the rest of the body before they are welded to the front wings. The straight edge that is sitting on the top of the pods is sighted across to the bulkhead to check that they are level with it. The vertical winding sticks are used to sight across to make sure that the front faces of the pods are in alignment.

during restoration to ensure that the doors are dimensionally correct.

To check that a panel or structure is sitting flat, or at the correct degree of twist, sit it on a flat surface such as a large bench, or use winding sticks to sight across. The winding sticks must be positioned parallel to each other to give a true sighting. When looking across the closer stick, the stick further away should be level with it.

Sighting across repair or replacement panels by eye when fitting to the bodyshell is used more than taking measurements, as there will be discrepancies dimensionally in sizes and shapes of panels, particularly on the complex double curvatures that are common on vehicle bodies. **Alignment is more important than measurement in assembling a bodyshell that needs to look square and symmetrical following restoration.**

JIGS

BODY ALIGNMENT JIG

Some form of body jig is necessary if carrying out major panel replacement to ensure that the body alignment is maintained, such as when replacing complete floor panels or sill sections on a bodyshell. The jig can be homemade to suit your vehicle, as manufactured jigs are expensive for an amateur to purchase and only likely to be used to restore one vehicle. The jig should be made from rigid sections of rectangular steel tube for strength; it is important that the jig does not distort during the restoration process. It should be rigid enough to withstand the structure of the bodyshell being clamped to it with no possibility of it being twisted or bent. Check the jig for parallelism, squareness and twist as you build it. Attachments for suspension, subframe and engine mounting points should be incorporated. The measurements for the positions of these points are sometimes available in original maintenance manuals. If this is not the case for your vehicle you may be able to find the information from your specific model classic car club or other enthusiasts.

It is worth fitting wheeled castors to the jig to allow it to be moved around the workshop. Make sure that these have a brake mechanism fitted so that the jig will not move unintentionally when it is being worked on.

ROLLOVER JIG

A rollover jig is particularly useful when welding on a vehicle bodyshell, as it is almost impossible to weld cleanly underneath a horizontal surface. The most convenient position in which to weld to achieve an effective weld, using the MIG or TIG process, is in a horizontal or downwards direction on top of a horizontal surface, or on a vertical surface – **not upside-down,** which is unsafe due to falling sparks and will not produce a satisfactory weld. A rollover jig allows you to position the bodyshell in the most convenient position for each repair being carried out. This needs to be a very robust frame that can support the weight of the bodyshell and should be balanced to allow easy adjustment of the angle at which the bodyshell is supported. Make sure that there is some form of locking device fitted which ensures that the body can't rotate unintentionally as it is being worked on. The height of the jig from the floor is an important consideration. This will be a compromise between having the body high enough to work on it comfortably and not too high that it can't be rotated inside the workshop.

BEFORE FITTING PANELS ON A VEHICLE

COMPETENCE

It is necessary to become competent in using the welding techniques as described in the previous chapters on practice pieces of scrap sheet metal before tackling repair work on a vehicle body. Once you are confident regarding your abilities, move on to welding in panels or patch repairs to the bodyshell. Even when fully competent, it is still wise to carry out practice on spare pieces of sheet metal, particularly if you haven't welded on that actual material for some time, before welding on a bodyshell.

Radius the corners of any repair panels or patches that are butt-welded into the bodyshell to ease fitting and finishing.

POSITIONING OF REPAIR JOINTS

Do not attempt to weld to metal that is badly corroded or too thin; fit a repair panel large enough to ensure that all corrosion or thinned areas are removed. Any repair joints are best placed on a curved section of the panel, which has more rigidity and so helps avoid distortion of the panel. Do not place a repair joint on a flat area of a panel unless it is absolutely necessary, as significant distortion will result. A common example of this is when repairing a door skin. The area of corrosion is usually around the bottom 150mm (6in) of the panel. However, it is best to fit a much larger panel that is joined on the top radius of the skin on or near a swage line, both of which add rigidity to the panel, so avoiding the excessive distortion that would result from welding on a flat or shallow curvature area of the panel.

To ease the fitting of a repair panel, radius the corners of the panel where possible, so that you are not welding into a sharp corner. Attempting to weld into a tight corner will tend to create gas inclusions at the point of the corner. It will also be more difficult to achieve a smooth finish to the panel's surface following welding as a greater degree of distortion will occur at the corner.

ORDER OF FITTING PANELS

It is best to follow the original manufacturer's order of assembly so as to ensure that the panels can be welded successfully, using the original joining methods if possible. It is worth spending time carrying out research to determine how the bodyshell was originally put together. If the correct order of assembly is not followed, it may be impossible to access some joints to weld them effectively.

It will occasionally be necessary to remove a panel from the bodyshell to repair it effectively, particularly where access to the back of the panel is restricted. Access is usually needed to hold a dolly behind the panel in order to hammer the joint smooth. It is also sometimes necessary to remove a panel in order to gain access to a corroded structure behind it that needs repair even when the outer panel is sound.

Remove all paint and any lead filler from the bodyshell around the repair joint to avoid contamination of the weld and toxic fumes being given off during welding.

Accuracy is key in aligning parts and in cutting the joint to ensure a good fit and finish to any repair. **Ensure that the repair panel is the correct shape and size before attempting to fit it.**

FITTING THE PANELS

Alignment of all panels is critical. All parts need to be clamped securely in place, as the expansion of metal caused by the heat produced in the welding process and shrinkage of the metal as it cools will move unclamped panels out of alignment.

FITTING A SPOT-WELDED PANEL

Initially tack the panel by placing spot welds at the end of the flanges and in the middle of the joint's length. Check that the panel alignment is correct before proceeding to place further spot welds in-between the initial welds. Allow time for the panel and the spot-welding machine to cool in-between welds. Follow the procedures laid out in Chapter Five.

FITTING A BUTT-WELDED PANEL

First, clamp the panel to the bodywork and mark around its perimeter with an indelible marker pen. This gives you a rough guide to then cut away the majority of the waste material, leave around 25mm inside of the marked line. This facilitates a better fit of the repair panel to the bodyshell. Reclamp the repair panel in place and mark around its perimeter accurately.

The repair panel is initially clamped in place and marked around its perimeter with an indelible ink marker pen to give an indication of how much waste material can be cut away before marking accurately.

Mark from the edge of the repair panel on to the wing using a straight-edged blade. An old hacksaw blade, cut off at an angle, is particularly useful for this.

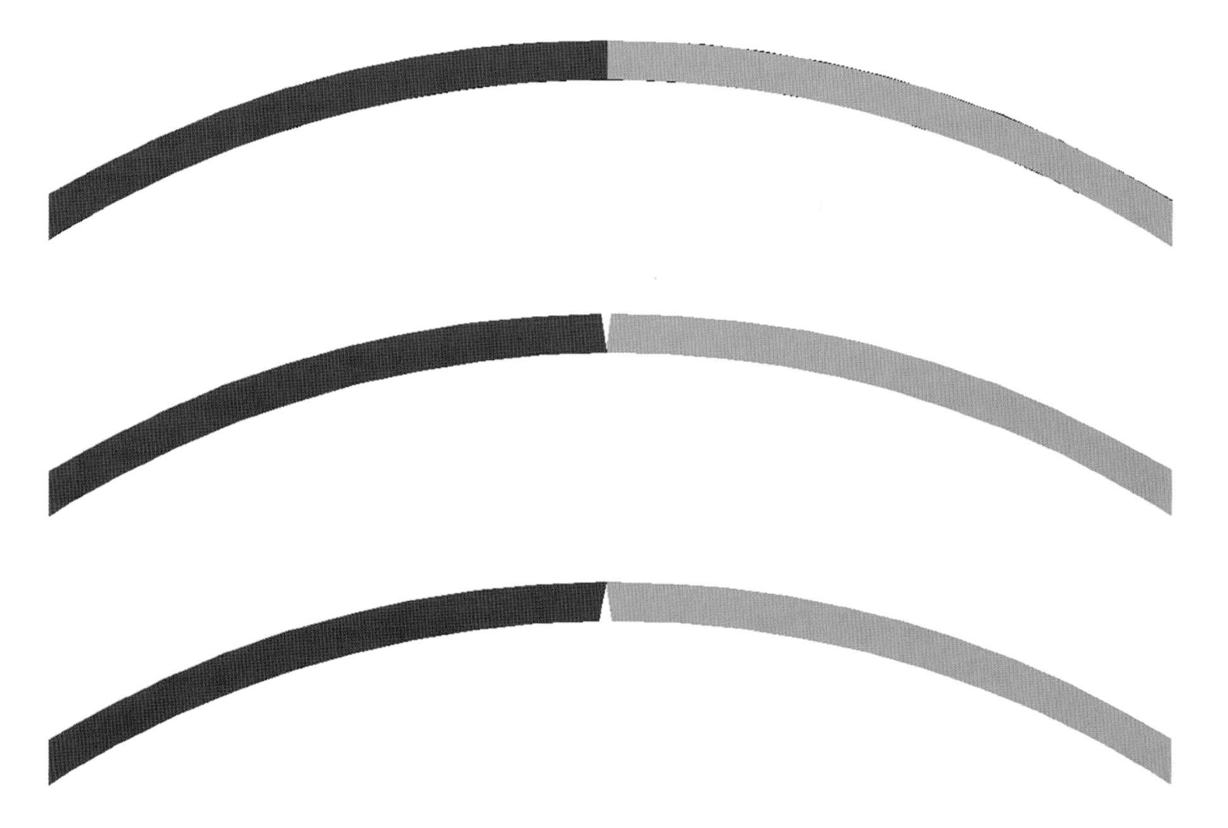

It is important to make sure that the edge of the panel is finished squarely to its surface to ensure that there is a tight joint for the weld.

Clamp the panel securely using locking pliers, or self-tapping screws if access is not available for clamps. Scribe around the perimeter of the repair panel using a straight-edged blade – an old hacksaw blade is good for this. Cut one end of the blade at an angle to create a sharp point. Keep the blade square to the surface of the panel whilst marking the joint.

Remove the repair panel and cut away the waste metal to the scribed line. This demands absolute precision, which is best achieved using hand tin snips. First cut away the majority of the waste material using a pneumatic hacksaw or other mechanical cutting device. This saves time and makes cutting with the tin snips easier, as cutting any width of metal with the snip is difficult and is likely to result in an inaccurate cut. Leave around 3–4mm for the final cut to be taken with the tin snips.

Finish by filing to the line, leaving a smooth edge for the weld. Make sure that the edges of the panels are filed square to the surface of the panel to give as tight a joint as possible. Any gap in the joint will increase the difficulty of welding it together cleanly. It is worth spending time at this stage to make sure that you have a tight joint before welding. **Do not deburr the edge of the panel.** This would leave a thinner edge, making it more difficult to weld on the joint.

TACK-WELDING A PANEL

Before beginning to tack-weld a panel in place, visually check that the welding machine and the welding

Panel tacked securely in place. Tacks placed around 20mm apart ensure that panels are kept level with each other as you progress.

Hammer the joint level and descale the tacks before running-in the full length of the joint.

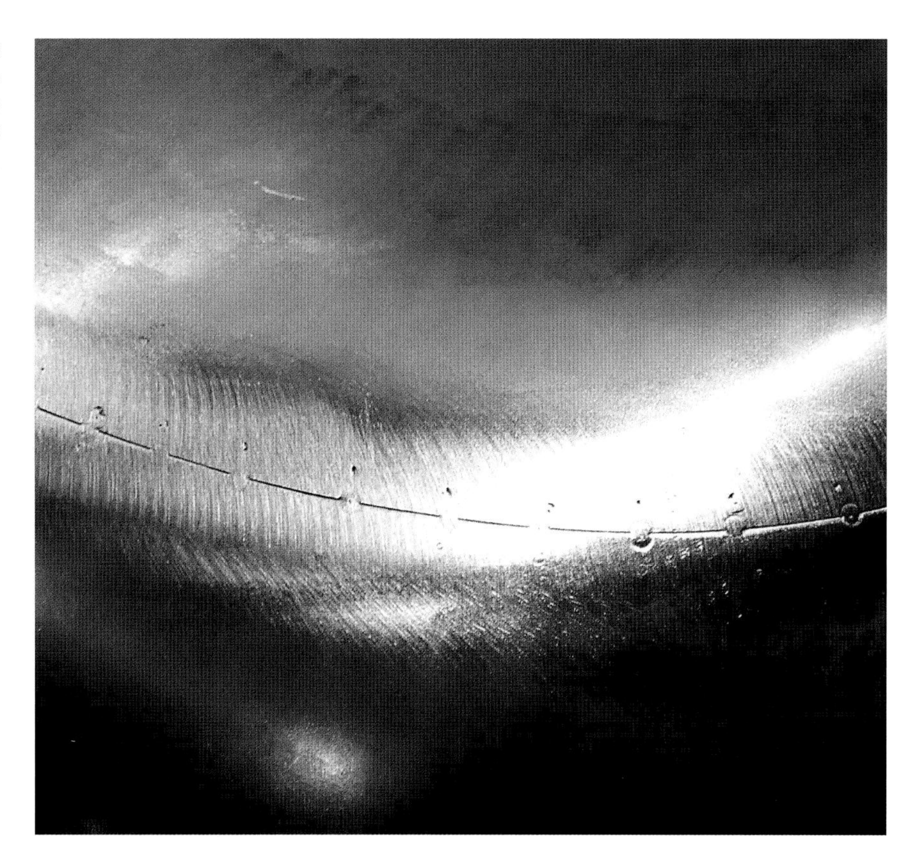

torch are set up correctly and that the gas is flowing at the correct rate. Test the welder on a scrap piece of sheet metal to ensure that the weld is clean and that the machine is running smoothly.

When welding along one long joint, clamp the panel at one end of the joint, holding the other end of the panel with your free hand, and place the first tack near the clamped end. Progress along the joint by moving away from this first tack, placing further tacks along the joint at 20mm intervals. Hammer the tacks lightly against a dolly held behind the joint using a small-headed hammer, in order to keep both panels level on the joint as you progress.

Once the tacking process is complete, hammer the tacks to level the panels. This also breaks the scale (surface oxide), making it easier to remove. Place a dolly directly behind the joint. Use a small-headed hammer to ensure accuracy and a dolly that is an accurate fit to the shape of the panel. Shape the dolly as necessary using a hand angle grinder so that it fits the contour of the panel in each direction. The shape of the dolly is more important than its finish, as this determines the finished shape of the panel.

Hammer lightly to avoid spreading the joint apart. To ensure accuracy when hammering, hold your elbow rigidly against the side of your body and move your hand by rotating it on your wrist joint. Maintain a loose grip on the hammer and allow it to bounce back after striking the joint. You should not feel any shock in your joints or tendons if done correctly.

RUNNING-IN THE JOINT

Remove the oxide scale from both sides of the tacked joint using a nylon sanding disc or abrasive hand pad before running-in the joint.

The joint can now be either fully welded in place with a continuous run of weld, or stitch-welded in place by making short runs of weld along the joint, gradually filling in the complete length of the joint. A continuous run of weld is best carried out where the repair is on a double-curvature shape, as the distortion from the welding process will have a minimal

Carry out a long a weld, as is comfortable. Descale the end of the weld run before restarting to avoid inclusion of oxide.

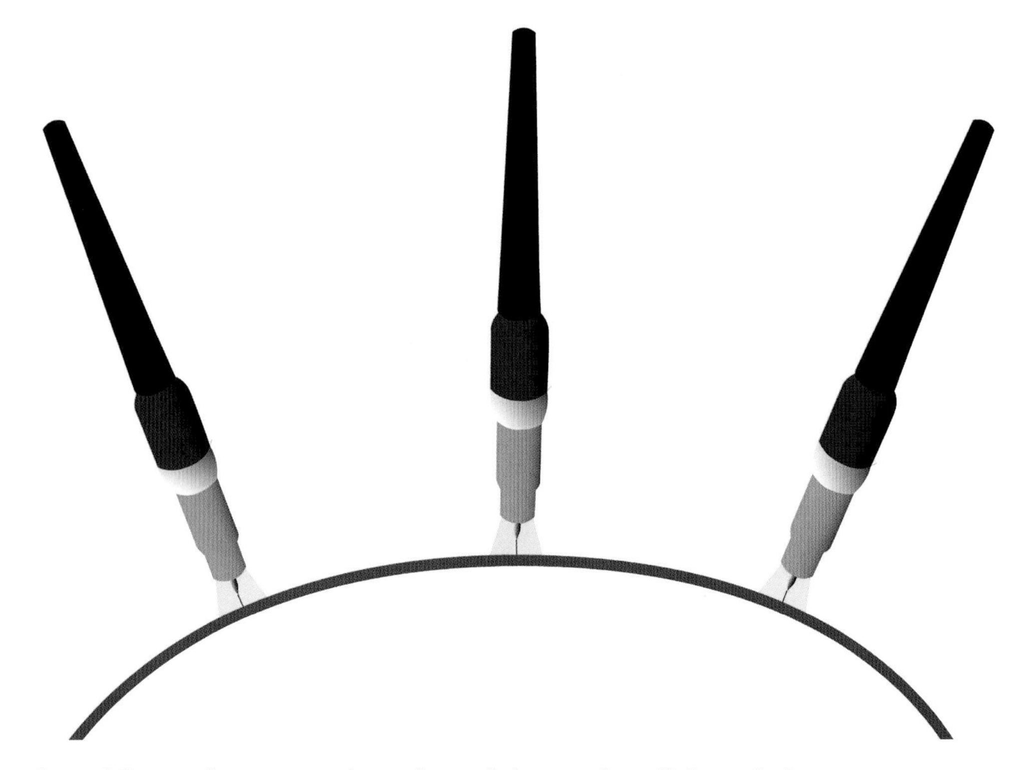

Keep the welding torch square to the surface of the panel at all times during welding to ensure that the welding gas is shielding the weld effectively.

effect because the shape of the panel will hold it rigid. The welded joint will also tend to peak up during the cooling of the metal, which gives a favourable shape to then grind off the excess material added if using the MIG process, and for hammering the joint level against a dolly. Descale the end of a run of weld before restarting a new section of weld so as to avoid any inclusion of surface oxide into the weld pool.

Use the stitch-welding technique where the repair is on or close to a shallow curvature-shaped panel. This technique, if done slowly by allowing time between welds, will minimize the distortion that is likely to occur on a shallow-shaped panel. If a heavy weld is carried out on a shallow curvature-shaped panel, the surface of the panel will tend to dip down as the metal cools and shrinks following the welding process. This makes grinding and finishing extremely difficult, making it impossible to achieve an acceptable finish to the shape of the panel, so should be avoided.

In all instances, keep the torch as square to the surface of the panel as possible. This means altering the angle of the torch to maintain an upright position relative to the surface shape of the panel as you progress.

FITTING A REPAIR PATCH ON A VEHICLE

MARKING THE JOINT

First, clean the area around the repair. Clamp the patch in place on the panel and mark around its perimeter with an indelible marker pen. Clean the area on and at least 50mm around the marked line, removing all paint, surface oxide and lead filler. Clamp

Clamp the repair patch in place and mark around its perimeter on to the wing, initially using an indelible ink marker pen, followed by a straight-edged blade.

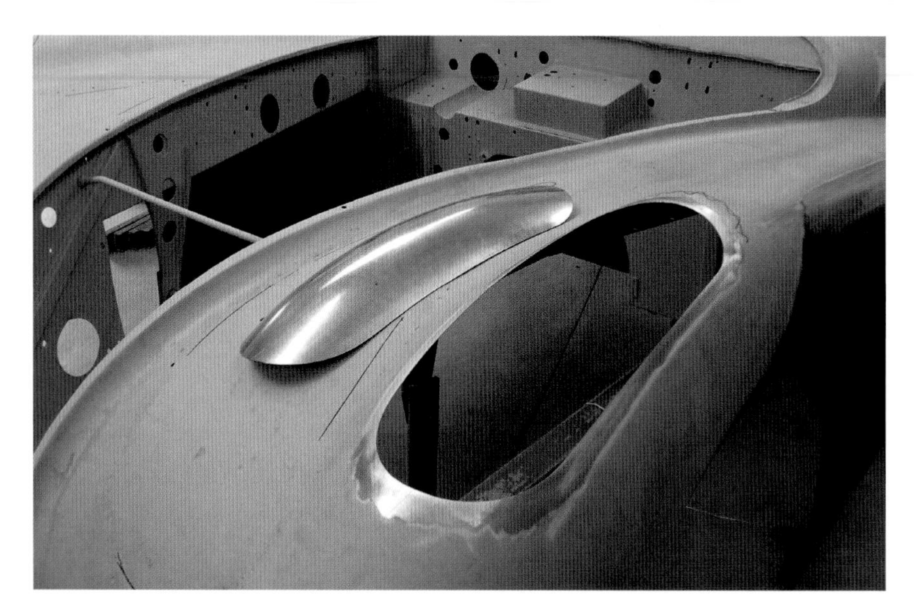

Cut to the line marked on the wing using tin snips. File the cut edge smooth and square to the surface of the panel to provide a tight butt joint ready for tacking the repair patch in place.

Tack the patch in place starting at opposite sides of the panel, then working in-between these initial tacks until enough tacks are in place to hold it securely.

the repair patch back in place on the body panel and mark around its edge with a sharp blade to give the cut line for the weld joint.

Cut the edge of the repair patch to the scribed line and file the edge smooth and square to its surface, ready for welding in the patch.

TACKING THE PATCH IN PLACE

When welding in a patch, begin by placing a tack on opposite sides of the patch, working progressively in-between these first tacks until enough tacks are in place to hold the repair securely. Ensure that the two panels are kept level as they are tacked together. The repair patch will usually be made of a slightly thicker metal than the original panel, which is likely to have been thinned by the effects of corrosion and blast cleaning. This means that you will need to bias the welding torch on to the new panel along the edge of the joint to avoid blowing holes in the original panel. The original panel is likely to heat up more quickly because it is thinner.

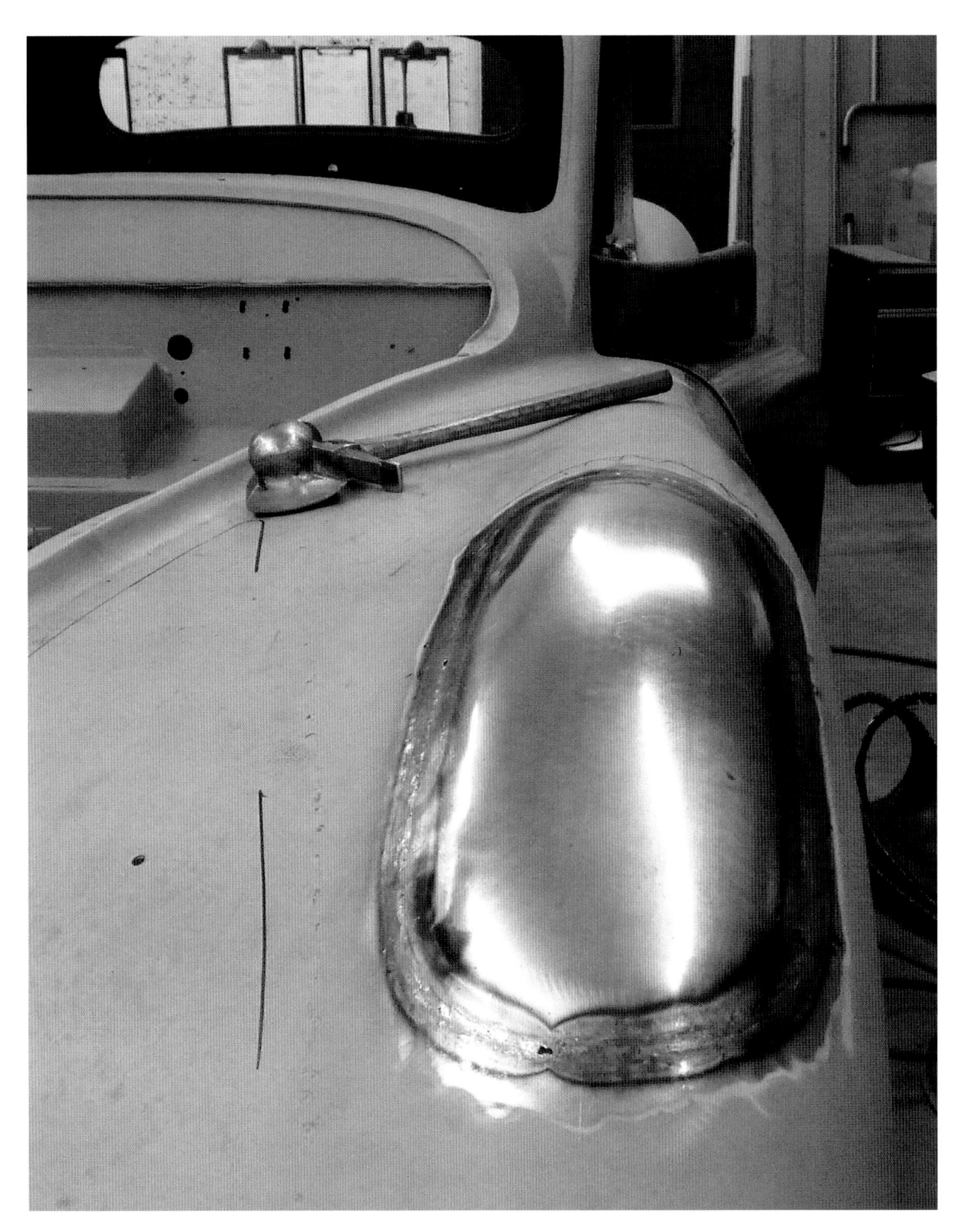

Once the panel is fully welded in place, hammer the joint to stretch the metal that has been shrunk by the welding process. Any holes created during running-in the weld can be repaired using the method described in Chapter Nine.

Descale the weld and planish with a large faced-flat hammer to smooth the contour of the panel. Lightly sand the joint and further planish if required to attain a smooth surface ready for paint finishing.

Clamp the bespoke repair patch in place to mark the cut line for the weld joint. The joint is marked with indelible marker prior to scribing.

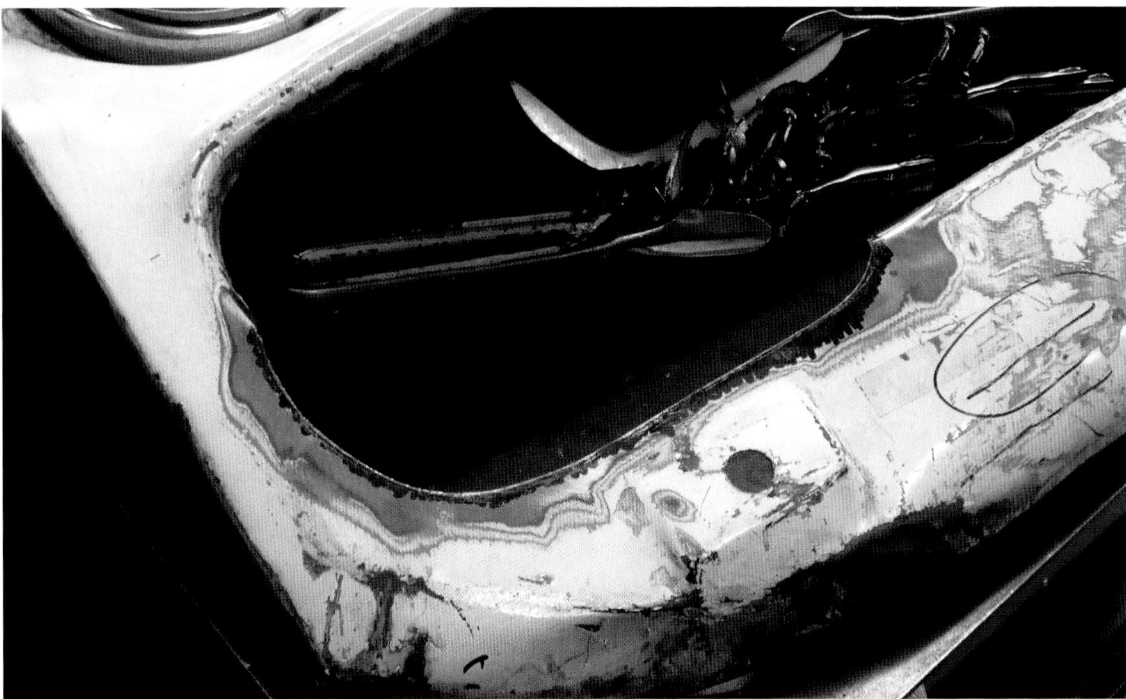
The panel is cut to the scribed line using tin snips, then the cut edge is filed smooth to provide a tight butt joint ready for tacking the repair patch in place.

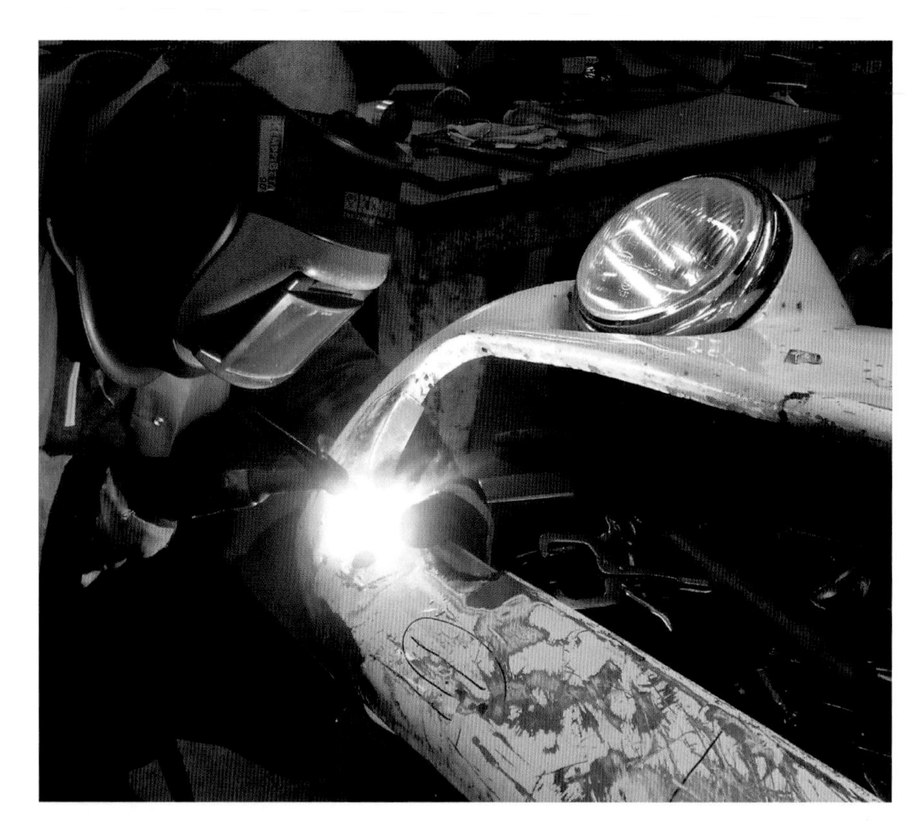

Tacking the repair panel in place.

Ensure that the panels are kept level on the joint as it is tacked in place. Use a small-headed hammer and dolly shaped to fit the contour of the panel.

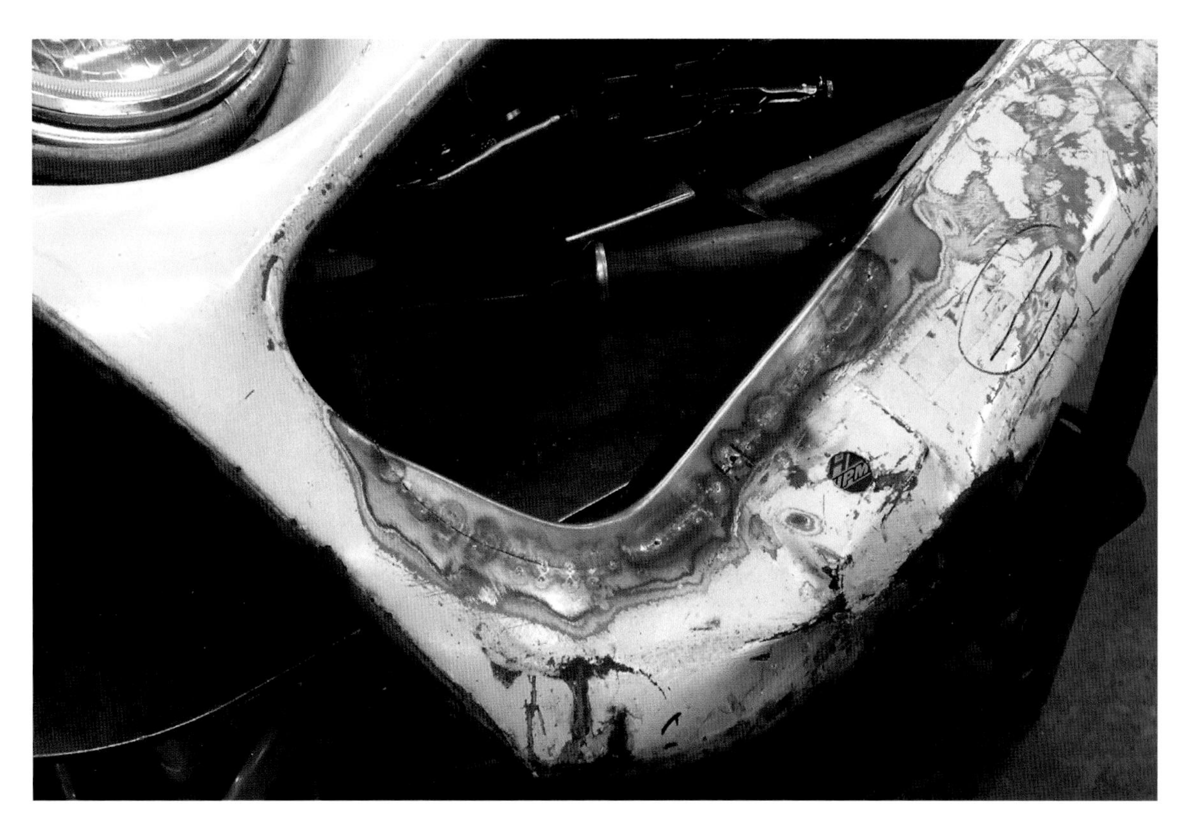

The patch is almost fully welded in place. There is always a likelihood of blowing holes in old panels that have been thinned by corrosion. Do not be overly concerned about this as they can be repaired later.

Using a nylon sanding disc, remove the oxide scale produced by the tacks. Hammer the joint to level both panels with a dolly held behind. Use a small-headed hammer to ensure accuracy and a dolly that is an accurate fit to the shape of the panel.

Fix the patch in place with a continuous run of weld. Keep the torch as square to the surface of the panel as possible; if the panel is curved, this means following the surface of the panel and turning the torch in an arc as you progress. Once the weld is complete, it is then hammered to stretch the metal that has been shrunk during the welding process so as to relieve any distortion created.

Before further planishing, descale the joint with a hammer and dolly to smooth the contour of the panel. Grind the surface of the panel with a fine abrasive and planish further if required to achieve a smooth finish to the contour of the surface of the panel ready for painting.

CHAPTER NINE

FINISHING WELDED JOINTS

Any welded joint will require a degree of work to achieve the desired finish prior to painting the panel. The first part of the finishing process is to rectify any defects in the weld. This is followed by a mixture of planishing (smoothing by hammering) and sanding on and around the repair to remove surface oxide and achieve a satisfactory standard of finish for paint preparation.

All metals, but particularly aluminium alloy, will be annealed (softened) on a welded joint due to the heat used in the welding process. It is important to harden the material by hammering or wheeling, which also levels and smooths the joint at the same time.

List of equipment

- electric hand drill and drill bits
- small-headed hammer
- planishing hammer
- flipper/body spoon
- dollies
- angle grinder
- power sander
- body file
- inspection gloves
- dust mask.

RECTIFYING DEFECTS

The two common defects are holes and inclusions in the weld joint. Holes are caused by: contamination in the joint as it is being fused; gaps due to a poorly cut joint; or too wide a weld that builds up a deficit of material, eventually making the weld blow through on the joint – this is caused by using too high an amperage or too slow a traverse speed when running-in the weld.

Inclusions are bubbles within the weld due to: insufficient shielding gas – because the gas flow rate is set too low, or is impaired by a kink in the cable; the torch is held too high from the surface of the weld; or the weld is too heavy (too high an amperage used), which results in an inclusion forming on the back of the weld where it is not shielded by the welding gas. Inclusions can also be created by welding over a surface oxide that has not been cleaned off prior to welding, or because foreign particles have been introduced into the weld, such as grinding dust, oil and so on. Both defects are repaired in the same way, by initially drilling through the burn hole or inclusion to create a precise, clean hole.

TIG METHOD

The drilled hole is filled by riveting in a plug of the same shape, size and material. The plug can be created quite simply using a hand punch. Use the smallest size necessary to fill the hole or inclusion. Any size of hole can be repaired using this method but the larger the plug, the more work required welding it into the joint. A larger plug will also create more distortion around the repair due to the heat spread from welding it in place. If you keep the hole size to 2.5mm ([3/32]in), it can be easily fused into the joint with one pass. A larger plug will require more passes, working around the edge of the plug, to fuse it successfully into place.

Rivet the plug tightly into the drilled hole using a small-headed hammer to ensure accuracy of the

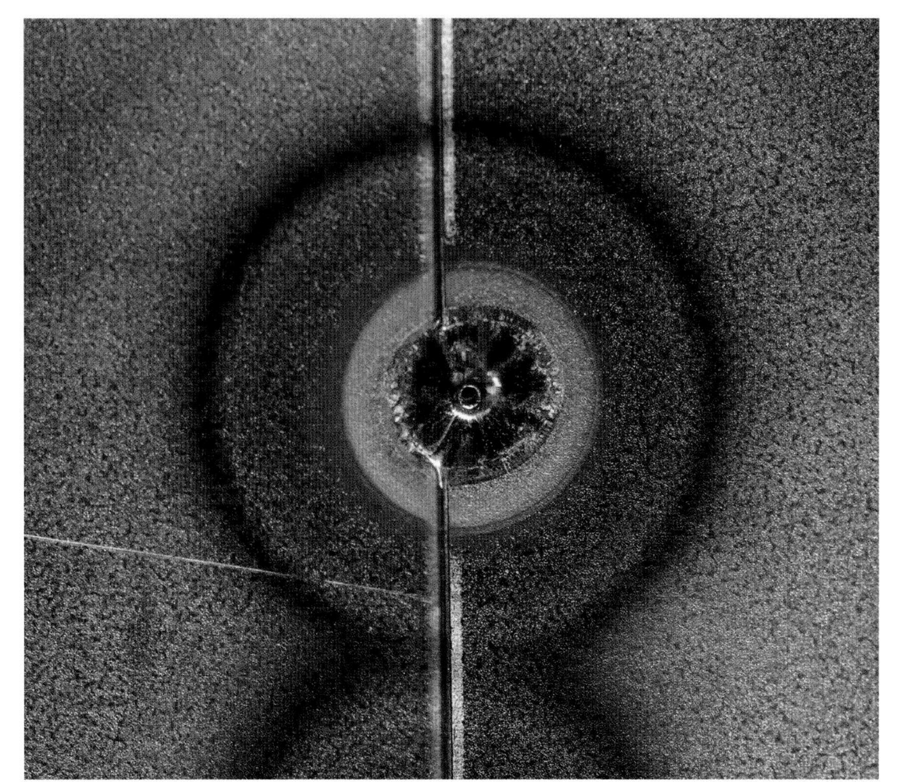

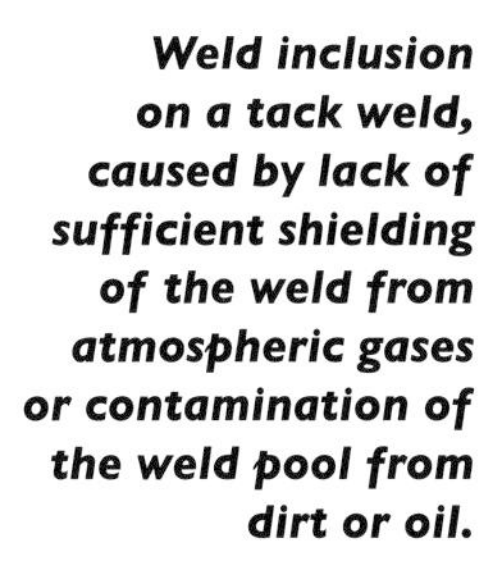

Weld inclusion on a tack weld, caused by lack of sufficient shielding of the weld from atmospheric gases or contamination of the weld pool from dirt or oil.

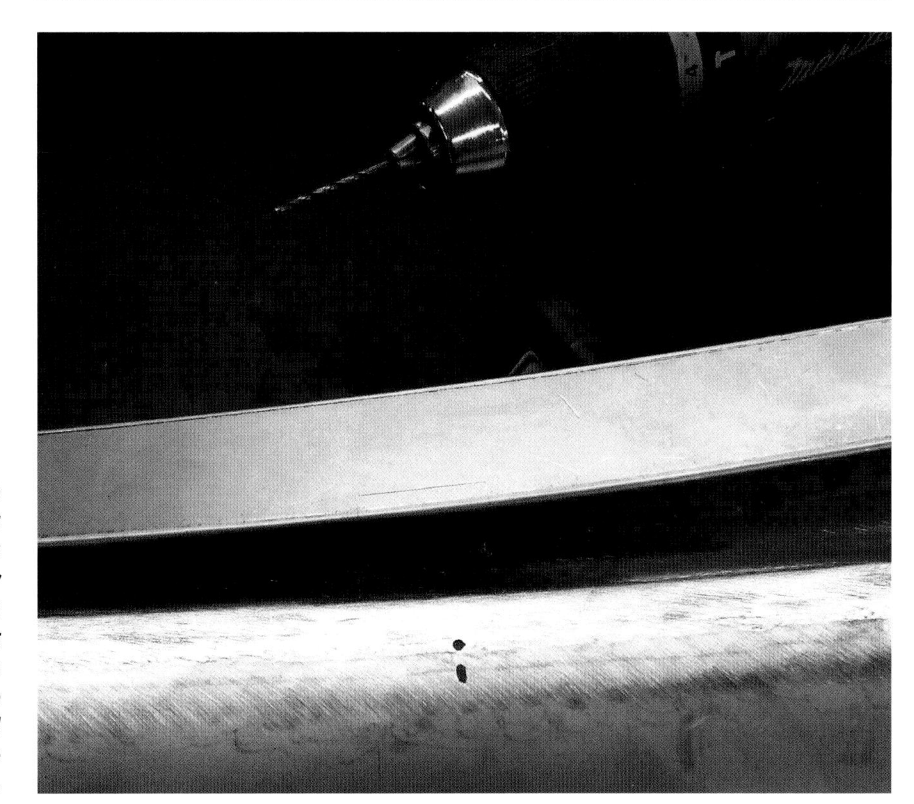

A clean, round hole is drilled in the joint to remove the area of an inclusion, or to create a uniform hole for repairing a blown hole in the joint. Use the minimum size drill needed to cover the defect.

A plug of sheet metal of the same diameter as the drilled hole is punched from the same thickness of material as the panel to fill the hole.

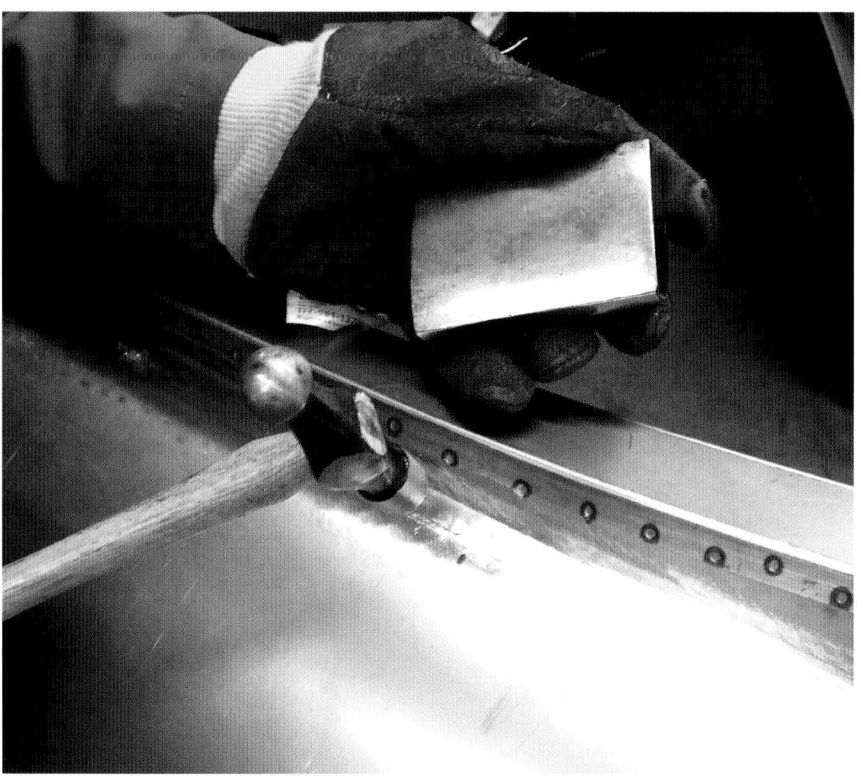

The plug is hammered into place using a small-headed hammer to ensure accuracy, with a dolly held behind the plug. The plug should be a tight fit following hammering, effectively riveting it in place.

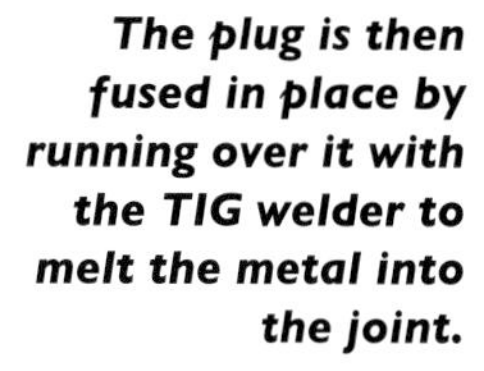

The plug is then fused in place by running over it with the TIG welder to melt the metal into the joint.

blows, with a suitably shaped dolly held behind the hole in the panel to hammer against.

Make sure that you clean the surface oxide from the metal between each pass, as welding on top of the oxide from the original weld will cause more inclusions, as any oxide will be melted into the weld. Do a short run of weld, starting before the plug and finishing after it, to integrate the plug into the joint. On a larger plug, weld around the edge of the plug to fuse it fully into place.

MIG METHOD

Use the plug-welding technique as described in Chapter Six, but instead of welding through to another panel hold a piece of thick copper plate behind the hole as you weld into the centre of the hole. This will leave a smooth back to the weld, as the molten steel filler wire will not stick to the copper backing but will fuse with the steel panel. Use the correct amperage needed to leave a slightly proud surface to the weld. Grind off any excess metal to leave the repair flush with the panel's surface.

STRETCHING THE WELDED JOINT

Following any repairs carried out to a vehicle body the completed welded joints are initially hammered to stretch the metal to remove any distortion created by the heat produced during the welding process. Be conscious of the need to stop hammering as soon as the tension is relieved, which is shown by the panel becoming smooth, as over-hammering the joint will create new distortion due to stretching the metal too much.

Use a small-headed hammer to ensure accuracy in hammering directly on to the weld joint. The welded material will be marginally thicker than the surrounding panel due to shrinkage over the width of the weld and the surrounding metal. The extra thickness of material created during the welding process provides a small amount of excess metal that can be ground or sanded back to remove small defects in the welded joint.

If possible, hammer the joint smooth while it is still hot; this means any slight distortion left will disappear as the metal cools and shrinks. Allow the

An area of a panel that has been stretched by damage from an impact or over-hammering can be shrunk by creating a hot spot with the TIG welding torch. Hammer the heated area flat whilst it is still hot.

metal to cool fully before checking for distortion. This can take a significant amount of time depending on the amount of heat introduced during welding. This action also hardens the metal on the weld, which will have been annealed during the welding process.

HEAT SHRINKING

Any overstretched areas can be shrunk back by creating a hot spot with the TIG torch. The metal expands around the hot spot and is flattened off by hammering lightly with a large diameter flat-headed hammer against a dolly of a suitable shape to fit the contours of the area on the panel being shrunk. This should be done immediately after creating a hot spot for the best effect. As the metal cools, the panel will pull flatter on and around the heated area. It is easy to overdo this so, make sure that you have allowed the panel to cool fully before deciding whether it is necessary to repeat the shrinking process. The area that has been heated and flattened off will usually require a small amount of stretching to relieve any excess shrinkage that is likely to occur on the heated area of the panel.

PLANISHING THE JOINT

Planishing is the term used for smoothing the contour of a panel with a flat-faced hammer by the action of making glancing blows across the surface of the panel with a suitably shaped dolly held behind the panel. This action flattens the panel to fit the shape of the dolly without stretching the surface of the metal, which would create more double-curvature.

Planish the joint smooth while it is still hot; any slight distortion left will disappear as the metal cools and the panel shrinks. Use a large diameter flat-faced, smooth-headed hammer, holding a dolly of the correct shape to fit the curvature of the panel behind the joint. Allow the metal to cool fully

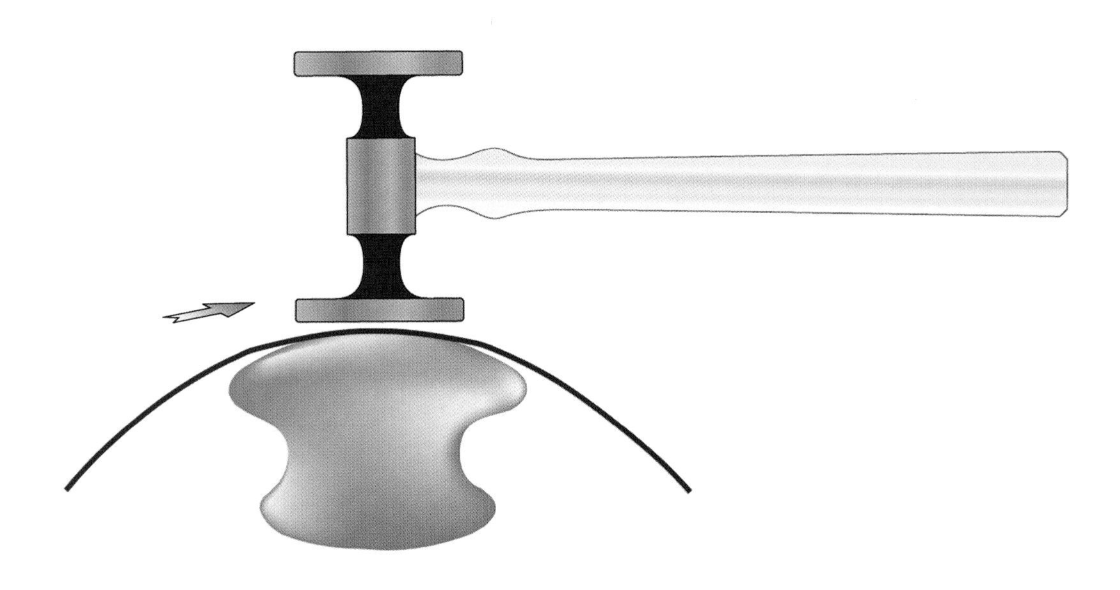

The aim of the planishing process is to smooth the contour of a panel without stretching the metal. This is achieved by using glancing blows with a hammer on the surface of the panel with a dolly held behind the panel.

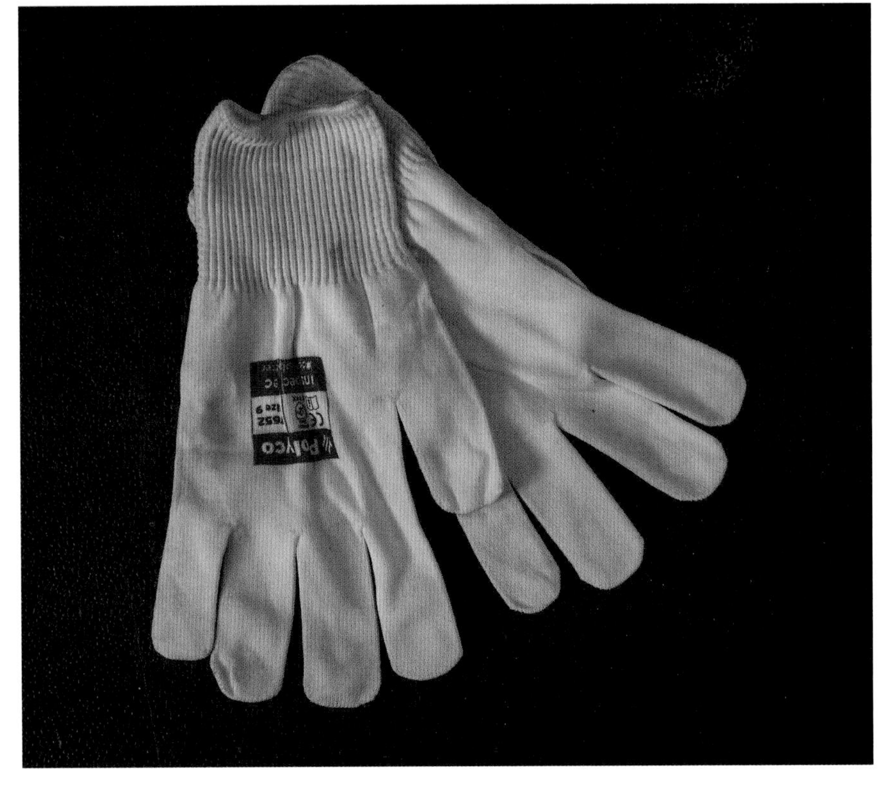

ABOVE: ***The welded joint should be left smooth enough following planishing to finish ready for paint without resorting to using fillers.***

Inspection gloves that are seamless, finely knitted nylon.

Wear a thin glove to feel the shape of a finished panel. Using a bare hand will not give an accurate sense of the contour of the panel, as the temperature and texture of the surface of the metal will override the feeling of shape.

before checking for distortion. On aluminium alloy panels use a flipper/body spoon to planish the joint rather than a hammer. Lightly sand the joint to highlight any high spots before further hammering and finally sanding to achieve the desired finish for paint preparation.

Feel the contour of the shape of a panel whilst wearing a thin glove to detect high or low spots. Wearing a glove insulates your hand from the temperature and surface texture of the panel and means that you are solely feeling the shape. Hold your complete hand flat against the panel while moving it swiftly along its surface; you are feeling for any movement across the pads of your fingers relative to the palm of your hand. Inspection gloves, which are made from finely knitted nylon and are seamless, are best used for this purpose.

GRINDING AND SANDING

The handheld angle grinder is one of the most useful power tools for grinding and sanding car bodywork when finishing welded joints. Various types of grinding and sanding discs and pads are available

An angle grinder will remove metal in the most efficient way, but care should be taken not to remove too much material, particularly on thin sheet metal. Oversanding also quickly produces a build-up of heat on a thin sheet-metal panel that may cause problems with distortion as the metal cools and shrinks. Apply the minimum amount of pressure needed on any sanding machine to avoid a build-up of heat in the panel. Use a minimum of 80grit grade of sanding disc; better still a fleece flap wheel that removes the minimal amount of metal and avoids any grinding marks that may be caused by digging in the edge of a card-backed disc or solid grinding wheel. Be aware that any grinding or sanding process can create a significant amount of heat due to the friction created. It is important to avoid a build-up of heat, as this can create distortion in the panel as the metal cools and shrinks.

A nylon woven disc is useful for removing oxide from the surface of a panel without grinding away the metal. A fleece flap disc will grind a small amount of metal, leaving a smooth finish to the surface of the panel.

Use a coarse fleece flap disc to remove a minimal amount of material on an aluminium welded joint. Finish with a medium-grade or fine-grade flap disc.

On an aluminium alloy panel, the weld is finished in a similar way to steel. Use a coarse fleece flap disc to remove the minimal amount of material to leave a smooth finish.

PREPARING FOR PAINT

CHEMICAL CLEANERS

A good general cleaner for use on most metals is phosphoric acid. This can be bought from a car accessory shop in the form of Jenolite. The acid removes blue weld scale and any oxide from the surface of a panel and provides an excellent key for paint finishing.

AVOIDING FUTURE CORROSION

Following any welding process, it is important to protect the surface of the metal as soon as possible, as it is left in a clean, raw state that is highly prone to corrosion. Do not leave welded panels untreated as they will quickly corrode on the surface, particularly

A restored bodyshell that has been painted with a zinc-rich primer to protect panels from surface corrosion before it is sent for final paint finishing. No body filler has been applied.

if left in a humid environment. Use a phosphoric acid such as Jenolite to clean the surface of the panels and apply a zinc primer when the acid has dried (usually after thirty minutes). It is not necessary to wash off the acid with water. If panels are to be left for a significant amount of time it is best to spray the surface with a thin layer of gloss paint on top of the primer. This will seal the primer, which is porous, to avoid any ingress of moisture or contaminants.

GLOSSARY

Age hardening Effects of aging on certain aluminium alloys that result in a hardening of the material.
Alloy Mixture of a metal with another metal or element to create more strength in the material.
Annealing Recrystallization of a metal that softens it, making it more workable.

Coachbuilder Maker of wooden-framed vehicle bodies.

Dolly A steel block that is shaped to fit the contour of a panel, used to hammer against.
Double-curvature Curved in both planes, as in a bowl or saddle shape.
Ductile Ability of a metal to be drawn during cold working without fracturing.

Flanged Edge of a panel that is turned up at an angle.
Flux Compound used to prevent oxidization forming on the surface of a heated metal.

Malleable Ability of a metal to be compressed during cold working without fracturing.
Monocoque Single self-supporting structure, such as used on a vehicle without a chassis.

Planishing Action of hammering, using a glancing blow, to smooth the contour of a panel.

Tensile strength The degree of stress a metal can withstand before fracturing.

Weld pool Area of molten metal created by the heat from the welding process.
Wheeling Forming or smoothing a double-curvature sheet-metal panel using a wheeling machine. The machine compresses and creates double-curvature in sheet metal.
Work hardening Effects of working metal, by hammering or rolling, which results in a hardening of the material.

FURTHER READING

Porter, Lindsay *Classic Cars Restoration Manual* (Haynes Publishing, 1994)
Robinson, A., *The Repair of Vehicle Bodies* (Butterworth-Heinemann, 1993)
Smith, Bob *How to Restore Sheet Metal Bodywork* (Osprey Publishing, 1984)
Smith, F.J.M., *Fundamentals of Fabrication and Welding Engineering* (Longman Scientific & Technical, 1993)
Walker, P.M.B., *Materials Science and Technology Dictionary* (Chambers Harrap, 1993)

INDEX

RELATED TITLES FROM CROWOOD

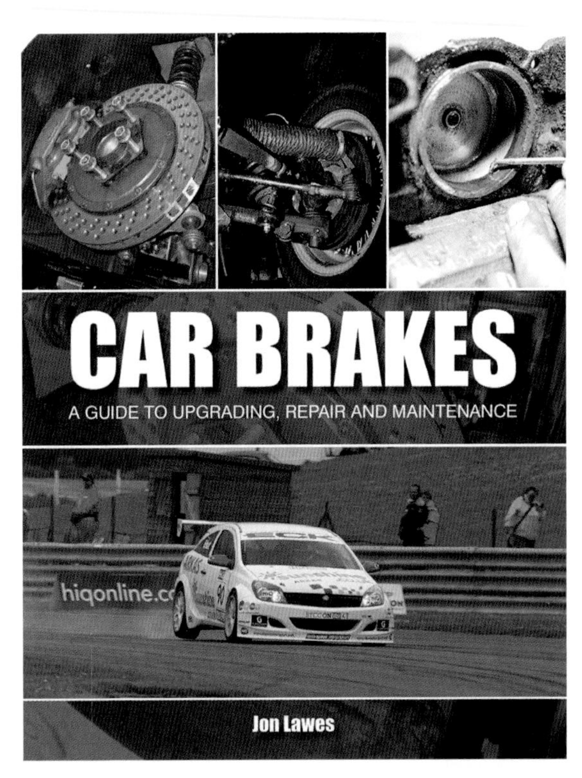

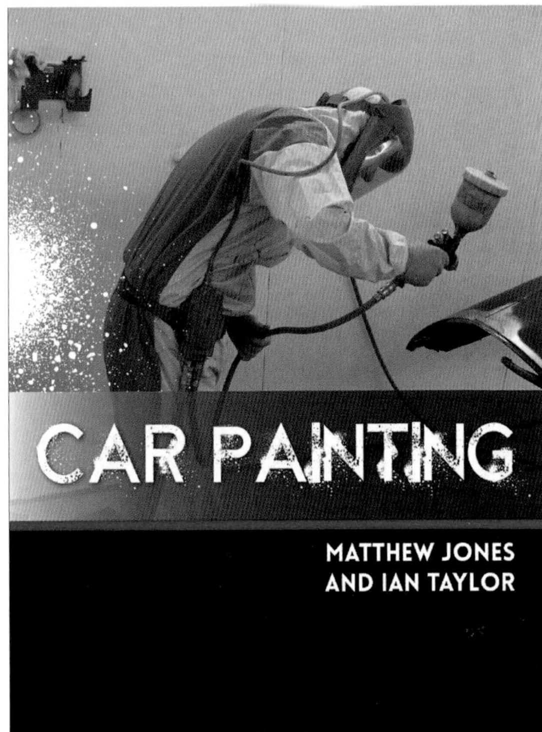

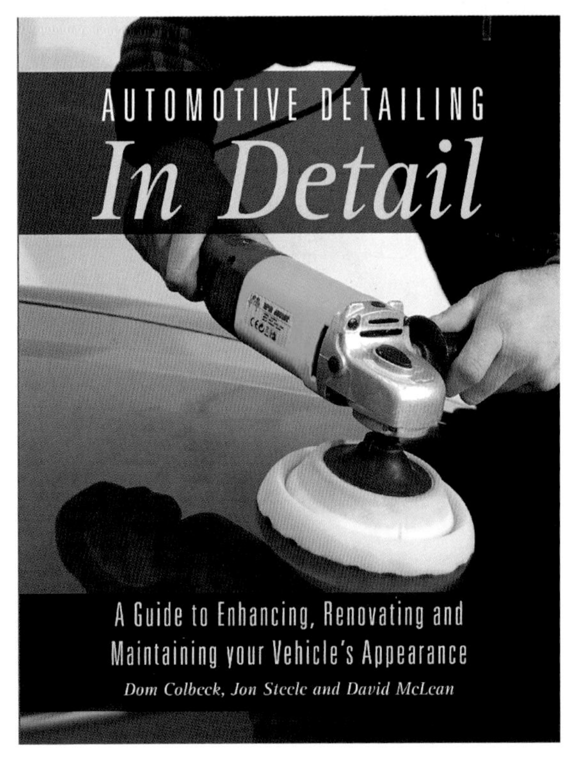